Guide to the
Umbral Calculus
A Different Mathematical Language

Guide to the
Umbral Calculus
A Different Mathematical Language

Silvia Licciardi
ENEA Research Center, Frascati, Italy

Giuseppe Dattoli
ENEA Research Center, Frascati, Italy &
University of Rome La Sapienza, Italy

World Scientific

NEW JERSEY · LONDON · SINGAPORE · BEIJING · SHANGHAI · HONG KONG · TAIPEI · CHENNAI · TOKYO

Published by

World Scientific Publishing Co. Pte. Ltd.

5 Toh Tuck Link, Singapore 596224

USA office: 27 Warren Street, Suite 401-402, Hackensack, NJ 07601

UK office: 57 Shelton Street, Covent Garden, London WC2H 9HE

British Library Cataloguing-in-Publication Data
A catalogue record for this book is available from the British Library.

GUIDE TO THE UMBRAL CALCULUS
A Different Mathematical Language

ISBN 978-981-125-532-8 (hardcover)
ISBN 978-981-125-533-5 (ebook for institutions)
ISBN 978-981-125-534-2 (ebook for individuals)

For any available supplementary material, please visit
https://www.worldscientific.com/worldscibooks/10.1142/12804#t=suppl

Desk Editor: Nur Syarfeena Binte Mohd Fauzi

Typeset by Stallion Press
Email: enquiries@stallionpress.com

"A Maria Corredentrice"
Silvia

"To my mother, Ada"
Pino

Contents

Preface

"Mathematics is the language through which God describes the World" (Galileo Galilei)

It is indeed a universal language shared by all peoples and which does not know the wear and tear of time. The mathematician does not invent anything but, at the most, discovers and it is immediately evident, in fact, that when a new formula is decrypted, it is instantly valid as true since before its discovery. However, there is a merit that the human being has, to employ his ingenuity and his passion to create new symbolisms and techniques so that the language is more and more usable and fruitful for everyone.

Umbral Calculus is only the latest in the chronological order of this set of writing and method. Although not widely known as it should be, it is developing rapidly thanks to its applicability to every branch of Pure and Applied Mathematics. It uses many transversal skills and is flexible, intuitive, ingenious in its apparent simplicity and, what is not indifferent, fun.

It reflects my way of living Mathematics and Life, using every available resource to get to the solution of a problem in the best way. I thank my illustrious colleague and co-author, Dr. Giuseppe Dattoli, for introducing me to the Umbral Calculus, during my studies and research. He, owing to his incredibly versatile knowledge of Mathematics and Physics, has made a unique contribution to this branch never developed before, which is making its way, not among a few oppositions, due to the undeniable effectiveness. While referring to the cardinal principles of Heaviside and the symbolism of Roman and Rota, Dr. Dattoli completely rewrote their meaning and use.

I hope that readers will enjoy using Umbral Calculus as much as we have enjoyed discussing it. Future implications are difficult to establish now, this is the beauty of Research.

"Ai posteri l'ardua sentenza"   (Alessandro Manzoni)

Silvia Licciardi, Ph.D.
Post-Doc Researcher at Enea Frascati Research Center
Palermo, September 14-th, 2021

In spite of the commonplace stating that Mathematics is not an opinion, it can at least be stated that there are different opinions about Mathematics. This is an example of different points of view between the Authors of the same book.

Mathematics is a language, the underlying rules represent the syntax and the symbols are the words. Any new formalism is just a different language that may provide advantages in terms of conciseness, computation, the possibility of connecting seemingly uncorrelated concepts within the same point of view or whatever...

This is however a naïve vision, underlying the idea that Mathematics makes progress when a new and more powerful language is formulated and reveals new elements bringing ahead the mathematical knowledge. The concepts we are trying to convey are largely undefined, so at least two questions arise:

What is a mathematical language?
What is mathematical knowledge?

Even though it may sound like blasphemy, the same naivety shines through the Hilbert statement[a] *(also reported as the epitaph on his tomb in Konisberg)*

"Wir Mussen wissen - Wir werden wissen"

which, like an ironic joke of destiny, was pronounced just one day before Gödel presented his thesis, containing the proof that any finite system of axioms, is not sufficient to prove every result in Mathematics. Gödel's Theorem decreed the end of the infancy of mathematical thought. To accomplish his program, he went much further with our statement, regarding the mathematical language, by making a distinction between Mathematics and meta-Mathematics.

---

[a]The English translation "We must know, we will know" loses its categorical strength.

The result of his work was that, within a certain mathematical theory based on a set of axioms, a given statement G cannot be demonstrated by the axioms themselves. Pushing further the above point, it might be concluded that G might be true or false, hence Mathematics is inconsistent or incomplete. The safe conclusion is that it is incomplete. Physical theories are supported by Mathematics and therefore self-referencing, like in Gödel's Theorem. One might therefore expect them to be either inconsistent or incomplete. This raises therefore serious doubts that a "complete" physical theory can be formulated as a finite number of principles. So Math is just the result of human activity and even if God pushed His pencil to formulate the world in mathematical terms, He made use of a formalism not accessible to us.

So coming back to the contents of this book, it is just an attempt of going a step further from the classical approach by Roman and Rota, the formalism we have developed is suited to treat a large number of problems in calculus and Special Function Theory.

I hope that the readers will enjoy the method as we enjoyed discovering it.

Prof. Giuseppe Dattoli
Former Senior Scientist at ENEA Frascati Research Center
Rome, September 16, 2021

Acknowledgments

This is a book on Umbral Calculus, merging different experiences, backgrounds and research aims of the Authors.

One (S.L.) has approached the problem as a Master's and Ph.D. student and later as researcher, after having gone through different works in Pure and Applied Mathematics.

The other (G.D.) went through a long process (for anagraphical reasons) which has brought him from high energy to laser Physics.

Both have therefore confronted and collaborated with a large number of colleagues who have opened their mind and contributed to the intellectual growth which led them to appreciate algebraic and symbolic methods as a powerful tool of computation. We would like to express our sincere appreciation to all of them.

A particular mention goes to Emanuele Di Palma, a rigorist who always looked at our methods with some concern, but generously helped us to check computations and to correct misconceptions.

It is also a pleasure to thank Prof. Antonio Siconolfi (Rome University La Sapienza), for his interest and the encouragements and Emeritus Distinguished Prof. John Rider Klauder (University of Florida) for his appreciation about the present book and his esteem, from a scientist of so much fame, it is a true honor.

The Authors

We would also like to express different thanks to people dear to us.

I express my very deep gratitude to my family for providing me with unfailing support and continuous encouragement and patience through the process of researching and writing. This accomplishment would not have been possible without them. Thanks to you all, to my husband Sandro and especially to God.

Silvia Licciardi

I like to mention the late Amalia Torre for the high quality work she produced during her life-time and for the rigor and deep thinking she put in her work.

Giuseppe Dattoli

Introduction

In the development of *Mathematical Methods*, an important step was taken by *John Blissard* in the second half of the 19th century. In his essay of 1861, *Theory of Generic Equations* [29], he established the rules which allowed the possibility of viewing completely different functions as realization of the same abstract entity. The Blissard's point of view inspired other researches. In particular those of *Edouard Lucas* [121], who referred to this methodology as *Umbral Calculus* and showed its usefulness for a number of applications. In his seminal piecework *Theorie nouvelle des nombres de Bernoulli et d'Euler* of 1877, Lucas pointed out that the method proposed by Blissard allowed the treatment of Bernoulli and Euler polynomials as ordinary monomials. Analogous ideas have been reshuffled from time to time. *Johan Frederik Steffensen* [162] introduced the concept of "poweroid" in 1940s and, more recently, the formalism of quasi-monomiality [47] has been established by an author of the present guide, *Giuseppe Dattoli*, in collaboration with the late Dr. Amalia Torre.

Within this last framework, it has been shown that most of the special polynomials can be viewed as ordinary monomials under the action of explicitly defined multiplicative and derivative operators. Some controversy arose at the beginning about the priority of the discovery, whose paternity was initially attributed to Lucas, but, first *James Whitbread Lee Glaisher* [97] and successively, in a more effective form, Eric Temple Bell [24] established the historical truth.

The presentation of Blissard's method by Bell in [24] is extremely effective and we summarize it in the following recapitulation.

We start by recalling that x_n, $n = 0, 1, \ldots$ denotes a sequence of real and complex numbers. A letter x without any suffix (which we will indicate here with $\bar{x}$) is the *representative* or *Umbra*[a] of the sequence x_n. Two Umbrae $\bar{x}, \bar{y}$ are equal if the corresponding sequences are equal. The umbra equality, as better discussed in the following, can be viewed as a property of Abstract Algebra (equality is symmetric, transitive and reflexive).

The starting point of the Blissard "revolution" was the observation that if x_n is written as the formal power of the corresponding umbra, namely $\bar{x}^n$, pushing forward this hint we find

$$x_{n+m} = \bar{x}^{n+m} = \bar{x}^{m+n}, \quad \forall n, m \in \mathbb{N},$$

$$(x+y)_n = (\overline{x+y})^n = \sum_{s=0}^{n} \binom{n}{s} \bar{x}^{n-s} \bar{y}^s = \sum_{s=0}^{n} \binom{n}{s} x_{n-s} \, y_s. \tag{1}$$

According to Bell, this is "the germ of the entire calculus".

The umbral "point of view" developed in the original researches by Blissard, Lucas and Bell lacked of a rigorous mathematical foundation, which came in the second half of the 20th century when Rota and Roman [149–151] framed the relevant concepts within the context of functional calculus, as we will discuss in the forthcoming chapters.

A different flavor of calculus has been developed more recently, it unifies old (Blissard, etc.) and new (Rota, Roman, etc.) points of view and provides a further step, allowing a significant simplification of

---

[a] Umbra is the latin term for "shadow". We will comment further on the relevant meaning inherent in the methods of umbral nature which, once applied to specific problems, regarding e.g., computations involving combinations of special functions, allow considerable simplifications.

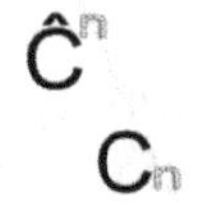

$$\hat{C}^n$$

$$C_n$$

Figure 1: Origins of the term Umbra according to Rota and Roman.

the computations underlying theory and practice of special functions and polynomials.[b]

The method we are proposing here goes beyond the common conception of Umbral Calculus, since it combines different technicalities, some of them borrowed from the *operational methods*, which yield the formulation of an albeit not fully rigorous, but certainly powerful *different mathematical language*.

Our attempt has been that of formulating a new tool embedding symbolic and umbral methods to be exploited in analytical or numerical computations involving, e.g., integrals, ordinary and partial differential equations, special functions and solutions to physical problems. The pivotal element of this strategy is the concept itself of "umbral image", which is the key element to establish the rules to replace higher transcendental functions in terms of elementary functions and the criteria to take advantage from such a replacement.

As previously noted, the term Umbra, even though introduced in an absolutely informal way, is the key-note to replace a series of the type

$$\sum_{n=0}^{\infty} c_n \frac{x^n}{n!},\tag{2}$$

representing a certain function $f(x)$ (with its x domain), with the formal exponential series

$$\sum_{n=0}^{\infty} \hat{c}^n \frac{x^n}{n!} = e^{\hat{c}x}.\tag{3}$$

The "promotion" of the index n in c_n to the status of a power exponent of the operator $\hat{c}$, namely the umbral operator, is the essence of "umbra", since it is a kind of projection of one into the other. Even though we adopt the same starting point, the conception of umbra and of the associated technicalities developed in this book are different. We will see that the possibility of replacing a function

---

[b]In order to avoid confusion, we underscore that in the formalism we will develop in the book, the sum of umbral operators is not equal to the operator of the sum $\hat{x}_1 + \hat{x}_2 \neq x_1 \hat{+} x_2$.

by a conveniently chosen formal series expansion provides significant advantages, among which is that of treating special functions as the *umbral image* of elementary functions. It should be noted that, within this point of view, the umbal image is the "umbra of a function". We will prove, for example, that *Bessel Functions* are the umbral images of the *Gaussian*. Albeit an apparently sterile exercise, such a point of view offers a wealth of new perspectives, based on a wise combination of "umbra", both for the study of the properties of old and new special functions as well as for the introduction of novel computational methods, differential operational calculus and algebraic manipulations.

The method we will expose has two venerable roots, namely the umbral tradition and the *Heaviside* operational calculus [130] along with the methods introduced by the operationalists (*Sylvester, Boole, Glaisher, Crofton and Blissard* [24, 35, 50, 51]) of the second half of the 19th century. Going back to the seminal papers by *Oliver Heaviside* in 1887, we quote the statement [104, 130]

> *There is a universe of mathematics lying in between the complete differentiation and integration.*

The Heaviside breackthrough in the theory of electric circuits was that of finding a way to treat the resistance, the capacitor and the inductor the same way, by the introduction of a specific operator, treated as an ordinary algebraic quantity. Within the framework of the Heaviside operational calculus, the analisys of whatever complex electric network can be reduced to straightforward algebraic manipulations.

The formalism introduced by Heaviside raised serious concerns in the mathematical community. Initially, it was rejected as nonsense, till it became clear that the proper mathematical environment to be placed in framing was the Laplace transform theory.

The motivations for our investigatios originated in an apparently straightforward question, namely:

"If we establish a formal equivalence between two functions, viewed as reciprocal umbrae, can we use the properties of one of the two to identify the properties of the other?"

The analysis has involved the study of umbral equivalence between rational, transcendental and higher order transcendental functions. We have established the rules within an umbral algebraic context and have eventually found the link with the theory of the *Borel* transform [54].

We have therefore followed a path for the search for a common thread between special functions, the relevant integral representation, the differential equations they satisfy and their group theoretical interpretation, by embedding all the previously quoted features within the same umbral formalism.

Any mathematical theory, even though elegant and capable of unveiling links between seemingly distant fields of researches, is viewed as academic if not capable of providing not previously known results. The following chapters contain a wide account of the applications of the method, involving the Theory of Special Functions, the evaluation of integrals, the solution of evolutive Partial Differential Equations (including those exploiting derivatives of fractional order), the Theory of Asymptotic Expansions and so on. The attempt has been that of designing a book with a manifold usefulness, capable of providing an account of the Umbral "theories" along with the relevant suitability of solving old and new problems, with minimum computational effort.

The contents of the present monographh cover different aspects of Applied Mathematics. They span over topics not usually covered in standard university courses. We use indeed the Theory of Generalized Multi-variable and Multi-index Special Functions and Polynomials. They are duly introduced and explained, but an initial effort is required of the reader to master and handle the underlying conceptual and technical novelties. To aid this we prepared a number of exercises for each chapter which provide a necessary complement to the "theoretical" part. We have chosen to sacrifice (within acceptable limits) the rigor to the simplicity and to the clarity of exposition. We hope that our effort will be useful to mathematicians and to other researchers applying Math to different disciplines (physicists, economists, statisticians, etc.) and may serve as stimulus for further research.

Chapter 1

Operator Theory and Umbral Calculus

This chapter is aimed at introducing our view of "Umbral Calculus" and the underlying technicalities. The formalism we will develop largely relies upon the Theory of Special Functions. We, therefore, premise a few specific notions, complemented by a résumé of elements of operational calculus, which constitutes one of the backbones of the whole theoretical implant.

We will make large use, e.g., of the *Euler Gamma Function*, defined by the integral representation [2]

$$\Gamma(z) = \int_0^\infty e^{-\xi}\xi^{z-1}d\xi, \quad \mathrm{Re}(z) > 0. \tag{1.0.1}$$

and, more in general, by [2]

$$\Gamma(z) = \lim_{n\to\infty} \frac{n!\, n^z}{\prod_{r=0}^n (z+r)}, \quad z \in \{\mathbb{C} \smallsetminus \mathbb{Z}^-\}. \tag{1.0.2}$$

It is well known that this function, for natural values of the variable z, reduces to the ordinary factorial, it can accordingly be viewed as a generalization of such an operation.

The following well-known identities are easily derived:

$$\Gamma(n+1) = \int_0^\infty e^{-\xi}\xi^n d\xi = n!, \quad \forall n \in \mathbb{N}, \tag{1.0.3}$$

$$\Gamma\left(\frac{1}{2}\right) = \int_0^\infty e^{-\xi}\xi^{-\frac{1}{2}}d\xi = \sqrt{\pi}. \tag{1.0.4}$$

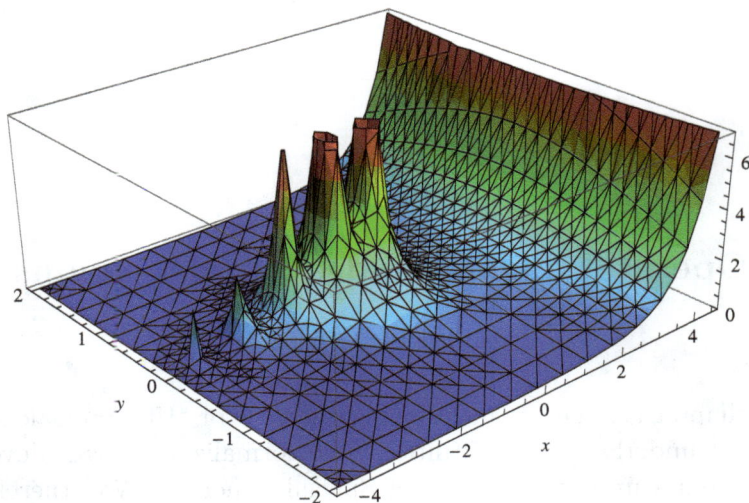

Figure 1.1:   Euler Gamma function $\Gamma(z)$ in the complex plane. The poles are present for $z \in \mathbb{Z}^-$.

Equation (1.0.3) is proved by repeated integration by parts and Eq. (1.0.4) is proved after setting $\xi = \mu^2$ and reducing the integral to a standard Gaussian integration.

The relevant plot in the complex plane is given in Fig. 1.1, which shows the poles for negative integer values of the argument.

Along with the Euler Gamma, another function (also introduced by Euler) will often be exploited here, namely the *Beta Function* (B-) defined in terms of the Gamma function as [2]

$$B(x,y) = \frac{\Gamma(x)\Gamma(y)}{\Gamma(x+y)} \tag{1.0.5}$$

which reads, for $\mathrm{Re}(x), \mathrm{Re}(y) > 0$,

$$B(x,y) = \int_0^1 t^{x-1}(1-t)^{y-1}dt. \tag{1.0.6}$$

Most of the special functions we will employ in the forthcoming sections are defined through a series expansion. Within this context, the use of the Gamma function is a pivotal tool specifying the terms of the series and the relevant properties of convergence. The Exercises at the end of the book complement and integrate the few defintions given in this section.

1.1 From Special Functions to their Umbral Images

We illustrate the "transition" from a *Special Function* to its umbral image by using a fairly straightforward example. The series (1.1.1) defines the *Bessel–Wright* (*BW*) function [179]

$$W_\beta^\alpha(x) = \sum_{r=0}^{\infty} \frac{x^r}{r!\,\Gamma(\alpha r + \beta + 1)}, \quad \forall x \in \mathbb{R}, \forall \alpha, \beta \in \mathbb{R}_0^+. \quad (1.1.1)$$

For $\alpha = 1$ and $x \to -x$, the above function is known as the *Tricomi-Bessel* (*TB*) function of order β [166]

$$C_\beta(x) = \sum_{r=0}^{\infty} \frac{(-x)^r}{r!\Gamma(r + \beta + 1)}, \quad \forall\, x,\, \beta \in \mathbb{R}. \quad (1.1.2)$$

The *Umbral Formalism*, which we use, relies on the simple assumption that functions of the type (1.1.1) can be treated as *ordinary exponential functions*, provided that we adopt the following notation for the *BW* function of 0-order.

Example 1.

$$W_0^\alpha(x) = e^{\hat{c}^\alpha x}\varphi_0, \quad \forall x, \alpha \in \mathbb{R}. \quad (1.1.3)$$

To prove this statement, we introduce the definition of "vacuum".

Definition 1. The function[a,b]

$$\varphi(\mu) := \varphi_\mu = \frac{1}{\Gamma(\mu + 1)}, \quad \forall \mu \in \mathbb{R}, \quad (1.1.4)$$

is called umbral **"vacuum"**.

This term, borrowed from physical language, is used to stress that **the action of the operators ĉ, raised to some power,**

---

[a]We remind that the inverse Γ-function is an entire function.
[b]The domain of μ can be extended to the field $\mathbb{C}$, but, in the present guide, our analysis is limited to the real cases.

is that of acting on an appropriate set of functions (in this case the Euler Gamma function) by "filling" the initial "state" $\varphi_0 = \frac{1}{\Gamma(1)}$.

Definition 2. We define the *Operator* $\hat{c}$, called *Umbral*,

$$\hat{c} = e^{\partial_z}, \tag{1.1.5}$$

the *vacuum shift operator*, where z is the domain's variable of the function on which the operator acts.

Theorem 1. *The umbral operator, $\hat{c}^\mu$, $\forall\, \mu \in \mathbb{R}$, is the action of the operator $\hat{c}$ on a vacuum φ_0 such that*

$$\hat{c}^\mu \varphi_0 := \varphi_\mu = \frac{1}{\Gamma(\mu + 1)}. \tag{1.1.6}$$

Proof. $\forall\, \mu \in \mathbb{R}$, applying Eqs. (1.1.5) and (1.1.4), we obtain

$$\hat{c}^\mu \varphi_0 = e^{\mu \partial_z} \varphi_z \,|_{z=0} = \left. \varphi_{z+\mu} \right|_{z=0} = \left. \frac{1}{\Gamma(z + \mu + 1)} \right|_{z=0} = \frac{1}{\Gamma(\mu + 1)}. \quad \Box$$

The umbral operator, defined according to the previous prescriptions, satisfies the following properties.

Properties 1. $\forall \mu, \nu \in \mathbb{R}$

$$\text{(i)} \quad \hat{c}^{\pm\mu} \hat{c}^\nu = \hat{c}^{\pm\mu+\nu}, \tag{1.1.7}$$

$$\text{(ii)} \quad \left(\hat{c}^{\pm\mu} \right)^\nu = \hat{c}^{\pm\mu\,\nu}. \tag{1.1.8}$$

Proof. $\forall\, \mu \in \mathbb{R}$

$$\text{(i)} \quad \hat{c}^\mu \hat{c}^\nu = e^{\mu \partial_z} e^{\nu \partial_z} = e^{(\mu+\nu)\partial_z} = \hat{c}^{\mu+\nu},$$

$$\text{(ii)} \quad \underbrace{\hat{c}^\mu \cdot \,\cdots\, \cdot \hat{c}^\mu}_{\nu} = \underbrace{e^{\mu \partial_x} \cdot \,\cdots\, \cdot e^{\mu \partial_x}}_{\nu} = e^{\mu\nu \partial_x} = \hat{c}^{\mu\,\nu}. \tag{1.1.9}$$

$$\Box$$

Remark 1. We underline that the action of the operator on the vacuum cannot be separated, it has to work in a *single action*.

We now have all the elements to prove Example 1.

Proof. We give a meaning to equation $W_0^\alpha(x) = e^{\hat{c}^\alpha x}\varphi_0$, $\forall x, \alpha \in \mathbb{R}$, by treating the r.h.s. as the exponential function of the operator $\hat{c}$ and thus, using an ordinary *MacLaurin* expansion, we end up with (see (1.1.8))

$$e^{\hat{c}^\alpha x}\varphi_0 = \sum_{r=0}^{\infty} \frac{\hat{c}^{\alpha\, r}}{r!} x^r \varphi_0. \tag{1.1.10}$$

The operator $\hat{c}$ acts on φ_0 only, *leaving x unaffected*, then **$\hat{c}$ and x commute** and we can cast the r.h.s. of Eq. (1.1.10) in the form

$$e^{\hat{c}^\alpha x}\varphi_0 = \sum_{r=0}^{\infty} \frac{x^r}{r!} \left(\hat{c}^{\alpha\, r}\varphi_0\right), \tag{1.1.11}$$

therefore, by applying the rule (1.1.6), we end up with

$$e^{\hat{c}^\alpha x}\varphi_0 = \sum_{r=0}^{\infty} \frac{x^r}{r!\,\Gamma(\alpha r + 1)} = W_0^\alpha(x), \quad \forall x, \alpha \in \mathbb{R}. \tag{1.1.12}$$

$\square$

It is also easily understood that, within such a formalism, the β-order BW function can be written as

$$W_\beta^\alpha(x) = \hat{c}^\beta e^{\hat{c}^\alpha x}\varphi_0, \quad \forall x \in \mathbb{R}, \forall \alpha, \beta \in \mathbb{R}_0^+. \tag{1.1.13}$$

In the forthcoming part of the book, we will take the freedom of treating *$\hat{c}$-like operator as having ordinary algebraic quantities* and, in the following section, we will see how the "associated" calculus finds its formal justification on the properties of the *Borel Transform* (BT).

1.1.1 *Borel Transform*

The theory of integral transforms is one of the fundamentals of the operational calculus. We therefore make a further step in this direction by providing a more rigorous environment to formulate the umbral technicalities established in the previous sections using as support the formalism underlying the theory of BT [55].

We show that, for the present purposes, the BT can be conveniently expressed in terms of Gamma function and of simple differential operators. Therefore, before proceeding further, we call the identity [55], $\forall \lambda \in \mathbb{R}, \forall x \in f's\ domain$,

$$e^{\lambda x \hat{D}_x} f(x) = f(e^\lambda x), \quad \hat{D}_x = \frac{\partial}{\partial x}, \tag{1.1.14}$$

whose proof is easily obtained after setting $x = e^\varsigma$ and noting that [18]

$$e^{\lambda x \hat{D}_x} f(x) = e^{\lambda \hat{D}_\varsigma} f(e^\varsigma) = f(e^\lambda e^\varsigma) = f(e^\lambda x). \tag{1.1.15}$$

As a consequence, the further identity[c] [18]

$$t^{x \hat{D}_x} f(x) = f(tx) \tag{1.1.16}$$

holds true. Furthermore, it is also important to stress that the monomial x^r is an eigenfunction of the operator $x\hat{D}_x$ in the sense that

$$(x\hat{D}_x)x^n = nx^n. \tag{1.1.17}$$

According to the previous discussion, the BT [91], expressed by the integral

$$f_B(x) = \int_0^\infty e^{-t} f(tx) dt, \tag{1.1.18}$$

can be recast in the operational form [55]

$$f_B(x) = \hat{B}\left(f(x)\right), \quad \hat{B} = \int_0^\infty e^{-t} t^{x \hat{D}_x} dt = \Gamma(x\hat{D}_x + 1). \tag{1.1.19}$$

A paradigmatic example, displaying how the $\hat{B}$ operator acts on a specific function, is provided by the 0-order TB function (Eq. (1.1.2)) [166], as shown in the following example.

---

[c]Or, in an alternative way, by setting $(\alpha)^{x \partial_x} = e^{\ln(\alpha) x \partial_x}$ and by making the change of variables $x = e^y$, we get $(\alpha)^{x \partial_x} f(x) = e^{\ln(\alpha) \partial_y} f(e^y) = f(e^{y+\ln(\alpha)})$, finally going back to the original variable we end up with Eq. (1.1.16).

Example 2.

$$C_0(x) = \sum_{r=0}^{\infty} (-1)^r \frac{x^r}{(r!)^2}, \quad \forall x \in \mathbb{R}. \tag{1.1.20}$$

The use of the identities (1.1.19) and (1.1.17) yields, $\forall x \in \mathbb{R}$,

$$\hat{B}(C_0(x)) = \Gamma(x\hat{D}_x + 1)(C_0(x)) = \sum_{r=0}^{\infty} (-1)^r \Gamma(r+1) \frac{x^r}{(r!)^2}$$

$$= e^{-x} = \sum_{r=0}^{\infty} (-1)^r \frac{x^r}{(r!)^2} \int_0^{\infty} e^{-t} t^r \, dt = \int_0^{\infty} e^{-t} C_0(tx) dt. \tag{1.1.21}$$

The $\hat{B}$ operator has evidently acted on the Bessel type function $C_0(x)$ by providing a kind of **downgrading** from higher transcendental function to the "simple" exponential.

Example 3. The successive application of the Borel operator to the same previous function produces the following result:

$$\hat{B}^2[C_0(x)] = \hat{B}[e^{-x}] = \sum_{r=0}^{\infty} (-1)^r \Gamma(r+1) \frac{x^r}{r!} = \frac{1}{1+x}, \quad |x| < 1. \tag{1.1.22}$$

Again, we note the same paradigm: the exponential function has been reduced to a rational function.

The further application of $\hat{B}$ yields a diverging series as shown in the following example.

Example 4.

$$\hat{B}^3[C_0(x)] = \sum_{r=0}^{\infty} (-1)^r r! x^r, \quad \forall x \in \mathbb{R}. \tag{1.1.23}$$

We have interchanged Borel operators and series summation without taking too much caution. In the case of Eq. (1.1.21), such a procedure is fully justified, in Eq. (1.1.22), the method is limited to the convergence region, while in the case of Eq. (1.1.23), the procedure is not justified since it gives rise to a diverging series. In the following, we will take some freedom in handling these problems and include in our treatment also the case of diverging series (see Chapter 7).

Since the repeated application of BT is associated with the Borel operator raised to some integer power, we extend the definition to a fractional BT and, more in general, to a real positive and negative power BT.

We introduce indeed the operator

$$\hat{B}_\alpha = \int_0^\infty e^{-t} t^{\alpha \, x \, \partial_x} dt = \Gamma(\alpha \, x \, \partial_x + 1), \quad \alpha \in \mathbb{R}^+, \qquad (1.1.24)$$

which will be referred to as the α-order BT.

Example 5. By using the *cylindrical Bessel Special Function* [18] (which will have a dedicated wider discussion in Chapter 4) $\forall x \in \mathbb{R}$

$$J_0(x) = \sum_{r=0}^\infty \frac{(-1)^r \left(\frac{x}{2}\right)^{2r}}{(r!)^2}, \qquad (1.1.25)$$

we find that the $\dfrac{1}{2}$-order BT applied to the zeroth-order Bessel yields

$$\hat{B}_{\frac{1}{2}}[J_0(x)] = \Gamma\left(\frac{1}{2} x \partial_x + 1\right) \sum_{r=0}^\infty \frac{(-1)^r}{(r!)^2} \left(\frac{x}{2}\right)^{2r} = \sum_{r=0}^\infty \frac{(-1)^r}{r!} \left(\frac{x}{2}\right)^{2r}$$

$$= e^{-\left(\frac{x}{2}\right)^2}. \qquad (1.1.26)$$

By assuming that $\alpha > 0$, there exists an operator $(\hat{B}^{(\alpha)})^{-1}$ such that

$$\left(\hat{B}_\alpha\right)^{-1} \hat{B}_\alpha = \hat{1}, \qquad \left(\hat{B}_\alpha\right)^{-1} = \frac{1}{\Gamma(\alpha x \partial_x + 1)}, \qquad (1.1.27)$$

then we can invert Eq. (1.1.26) and write

$$\left(\hat{B}_{\frac{1}{2}}\right)^{-1} \left[e^{-\left(\frac{x}{2}\right)^2}\right] = J_0(x). \qquad (1.1.28)$$

Observation 1. The extension of Eq. (1.1.27) to negative α yields

$$\hat{B}_{(-\alpha)} = \Gamma(-\alpha x \partial_x + 1) = \frac{1}{\Gamma(\alpha x \partial_x)} \frac{\pi}{sin(\alpha \pi x \partial_x)} \qquad (1.1.29)$$

and it is worth stressing that [55]

$$\hat{B}_{(-\alpha)} \neq \left[\hat{B}_{(\alpha)}\right]^{-1}. \qquad (1.1.30)$$

A definition of the inverse of the operator $\hat{B}_\alpha$ may be achieved through the use of the *Hankel* contour integral, namely

$$\frac{1}{\Gamma(z)} = -\frac{i}{2\pi} \int_C \frac{e^{-t}}{(-t)^z} dt, \quad |z| < 1, \qquad (1.1.31)$$

which can be exploited to write

$$\left(\hat{B}_\alpha\right)^{-1} f(x) = -\frac{i}{2\pi} \int_C \frac{e^{-t}}{t} f\left(\frac{x}{(-t)^\alpha}\right) dt. \qquad (1.1.32)$$

After the previous remarks, we can state the following theorem.

Theorem 2. *Let $f(x)$ be a function such where $\int_{-\infty}^{+\infty} f(x)dx = k, \forall k \in \mathbb{R}$, then*

$$\int_{-\infty}^{+\infty} \hat{B}_\alpha[f(x)]dx = k\,\Gamma(1-\alpha), \quad |\alpha| < 1. \qquad (1.1.33)$$

Proof. The proof is fairly straightforward by applying Eq. (1.1.18) and the variable change $t^\alpha x = \sigma$. $\forall k \in \mathbb{R}, |\alpha| < 1$. We find

$$\int_{-\infty}^{+\infty} \hat{B}^{(\alpha)}[f(x)]dx = \int_{-\infty}^{+\infty} \left(\int_0^{+\infty} e^{-t} f(t^\alpha x)dx\right) dt$$

$$= \int_{-\infty}^{+\infty} e^{-t} \left(\int_0^{+\infty} f(t^\alpha x)dx\right) dt$$

$$= \int_{-\infty}^{+\infty} e^{-t} t^{-\alpha} \left(\int_0^{+\infty} f(\sigma)d\sigma\right) dt$$

$$= \int_{-\infty}^{+\infty} f(\sigma) \left(\int_0^{+\infty} e^{-t} t^{-\alpha} dt\right) d\sigma = k\,\Gamma(1-\alpha).$$
$\square$

The same procedure can be exploited for cases involving the inverse transform.

These remarks provide a more sound basis for the formalism we are discussing and will be further developed and applied in the forthcoming parts. We will corroborate our conclusions using extensions of the concepts developed in this section.

1.2   Gaussian Function in Umbral Calculus

In the following, we exploit the properties of the Gaussian function in an umbral context and, in particular, we see that families of Special Functions, like Bessel functions, can be viewed as *Umbral representation* of the Gaussian itself. To this aim, it is worth call the following properties of the function e^{-x^2} [18].

We start with the well-known *Gaussian integral*

$$\int_{-\infty}^{\infty} e^{-x^2} dx = \sqrt{\pi} \tag{1.2.1}$$

and call the *Gauss–Weierstrass integral (GWI)*, which will often be exploited in the following,

$$\int_{-\infty}^{\infty} e^{-ax^2+bx} dx = \sqrt{\frac{\pi}{a}} \, e^{\frac{b^2}{4a}}, \quad \forall b \in \mathbb{R}, \forall a \in \mathbb{R}^+. \tag{1.2.2}$$

A particularly useful result, strictly related to (1.2.2), is given as follows by the *Gaussian integral identity (GII)*:

$$e^{-b^2} = \frac{1}{\sqrt{\pi}} \int_{-\infty}^{\infty} e^{-\xi^2 - 2\,i\,b\,\xi} \, d\xi, \quad \forall b \in \mathbb{R}. \tag{1.2.3}$$

Along with the Gaussian function, we introduce an associated family of special poynomials, which plays a crucial role for the topics treated in this and in the forthcoming chapters. The tight link between (ordinary) Hermite polynomials $H_n(x)$ and Gaussan functions is provided by the relevant definition through the *Rodrigues formula*[d] [3]

$$H_n(x) = (-1)^n e^{x^2} \left(\frac{d}{dx}\right)^n e^{-x^2} = (-1)^n n! \sum_{r=0}^{\lfloor \frac{n}{2} \rfloor} \frac{(-1)^r (2x)^{n-2r}}{r!(n-2r)!}.$$

$$(1.2.4)$$

In the following, we will use a generalized form of Hermite more useful for the present purposes.

Proposition 1. *Let*

$$H_n(\xi, \mu) = n! \sum_{r=0}^{\lfloor \frac{n}{2} \rfloor} \frac{\xi^{n-2r} \mu^r}{r!(n-2r)!}, \quad \forall \xi, \mu \in \mathbb{R}, \forall n \in \mathbb{N}, \qquad (1.2.5)$$

two variable polynomials often referred to as Hermite–Kampé de Fériét *polynomials, obtained by repetead derivatives of a Gaussian or also defined through the generating function* [5, 18]

$$\sum_{n=0}^{\infty} \frac{t^n}{n!} H_n(x, y) = e^{xt + yt^2}, \quad \forall x, y \in \mathbb{R}, \qquad (1.2.6)$$

then, it is possible to prove that $\forall a \in \mathbb{R}$,

$$\partial_x^n e^{-ax^2} = H_n(-2ax, -a)e^{-ax^2} = (-1)^n H_n(2ax, -a)e^{-ax^2}. \quad (1.2.7)$$

Proof. $\forall x, y \in \mathbb{R}$, using the shift operator (7.3.1) and Eq. (1.2.6) we get

$$\sum_{n=0}^{\infty} \frac{t^n}{n!} \partial_x^n e^{-ax^2} = e^{t\partial_x} e^{-ax^2} = e^{-a(x+t)^2} = \sum_{n=0}^{\infty} \frac{t^n}{n!} H_n(-2ax, -a)e^{-ax^2}$$

$$(1.2.8)$$

and, equating the same like power t terms, we end up with Eq. (1.2.7).

□

---

[d]For further details, see Chapter 7.

According to Eq. (1.2.7), it is also easily inferred that the ordinary Hermite are specified in terms of their two-variable counterpart, as shown in the following property (1.2.9).

Properties 2. $\forall x, y, a \in \mathbb{R}, \forall n \in \mathbb{N}$, *Hermite polynomials satisfy*[e,f] *the following*:

1.

$$H_n(x) = H_n(2x, -1) = 2^n H_n\left(x, -\frac{1}{4}\right), \tag{1.2.9}$$

2.

$$a^n H_n(x, y) = H_n(ax, a^2 y) \tag{1.2.10}$$

3.

$$H_n(x, y) = (-i)^n y^{\frac{n}{2}} H_n\left(\frac{i\,x}{2\sqrt{y}}\right), \tag{1.2.11}$$

4.

$$H_n(x, y) = (-y)^{n/2} H_n\left(\frac{x}{2\sqrt{-y}}\right), \tag{1.2.12}$$

5.

$$H_n(x, y) = y^{\frac{n}{2}} H_n\left(\frac{x}{\sqrt{y}}, 1\right), \tag{1.2.13}$$

6.

$$He_n(x) = H_n\left(x, -\frac{1}{2}\right), \tag{1.2.14}$$

7.

$$H_n(x) = 2^{\frac{n}{2}} He_n\left(\frac{x}{\sqrt{2}}\right), \tag{1.2.15}$$

---

[e] About property (1.2.14), we recall that $He_n(x)$ is a canonical form of one variable HP which uses $\mid y \mid = \frac{1}{2}$ in two variables $H_n(x, - \mid y \mid)$ and satisfies the generating function $\sum_{n=0}^{\infty} \frac{t^n}{n!} He_n(x) = e^{xt - \frac{t^2}{2}}$. They are called Hermite polynomials of Quantum Mechanics.

[f] We also remind the link between two- and one-variable HP (1.2.4) [18].

8.

$$H_n(x,y) = i^n (2y)^{\frac{n}{2}} He_n \left(\frac{x}{i\sqrt{2y}} \right). \qquad (1.2.16)$$

The exercises at the end of the book are a necessary complement to get confidence with the relevant properties and to master their use.

1.2.1 Umbral Bessel Function

To give a first idea of how powerful the umbral representation is, we consider the *cylindrical Bessel function of 0-order* (1.1.25) [18, 133] and its link with the TB functions.

Lemma 1. *By using the operator definition* (1.1.6) *and the property of* Γ*-function* (1.0.3)*, we find,* $\forall x \in \mathbb{R}$,

$$J_0(x) = \sum_{r=0}^{\infty} \frac{(-1)^r \left(\frac{x}{2}\right)^{2r}}{(r!)^2} = \sum_{r=0}^{\infty} \frac{(-1)^r \left(\frac{x}{2}\right)^{2r} \hat{c}^r}{r!} \varphi_0 = e^{-\hat{c}\left(\frac{x}{2}\right)^2} \varphi_0,$$

$$(1.2.17)$$

obtaining in this way a new formulation of Bessel function.

Lemma 2. *The 0-order TB* (1.1.2) *is expressible in the 0-order cylindrical Bessel as*

$$C_0(x) = J_0 \left(2\sqrt{x}\right), \quad \forall x \in \mathbb{R}. \qquad (1.2.18)$$

Proof. $\forall x \in \mathbb{R}$, by using (1.0.3) and algebraic manipulations,

$$C_0(x) = \sum_{r=0}^{\infty} \frac{(-x)^r}{r!\Gamma(r+1)} = \sum_{r=0}^{\infty} \frac{(-1)^r (\sqrt{x})^{2r}}{r!^2} = J_0 \left(2\sqrt{x}\right). \quad (1.2.19)$$

$\square$

Corollary 1. *We can write the umbral 0-order TB function*

$$C_0(x) = e^{-\hat{c}x} \varphi_0, \quad \forall x \in \mathbb{R}. \qquad (1.2.20)$$

Example 6. Using the *GWI* (1.2.2), the operator definition (1.1.6) and the Γ-function property (1.0.4), we obtain

$$\int_{-\infty}^{\infty} J_0(x)dx = \int_{-\infty}^{\infty} e^{-\hat{c}\left(\frac{x}{2}\right)^2} dx \; \varphi_0 = 2\sqrt{\frac{\pi}{\hat{c}}} \varphi_0 = 2\sqrt{\pi} \hat{c}^{-\frac{1}{2}} \varphi_0$$

$$= 2\sqrt{\pi} \frac{1}{\Gamma\left(-\frac{1}{2}+1\right)} = 2. \tag{1.2.21}$$

We end this section with the following lemma [52].

Lemma 3. *By the use of the GII (1.2.3), and Eq. (1.1.3), we obtain,* $\forall x \in \mathbb{R}$,

$$J_0(x) = e^{-\hat{c}\left(\frac{x}{2}\right)^2} \varphi_0 = \frac{1}{\sqrt{\pi}} \int_{-\infty}^{+\infty} e^{-\xi^2 - i\hat{c}^{\frac{1}{2}} x \, \xi} d\xi \; \varphi_0$$

$$= \frac{1}{\sqrt{\pi}} \int_{-\infty}^{+\infty} e^{-\xi^2} \left[e^{-i\hat{c}^{\frac{1}{2}} x \, \xi} \varphi_0 \right] d\xi$$

$$= \frac{1}{\sqrt{\pi}} \int_{-\infty}^{+\infty} e^{-\xi^2} W_0^{\frac{1}{2}}(-i \, x \, \xi) d\xi \tag{1.2.22}$$

which realizes a new integral representation of 0-order Bessel function in terms of 0-order BW function.

Now, the meaning of umbral image of a function is clear. Whenever two functions share the same formal series by means of the introduction of an appropriate umbral operator, they are reciprocal images.

The further consequence is also that, according to the previous discussion on Hermite polynomials, the following identity holds:

$$\left(\frac{d}{dx}\right)^n J_0(x) = \left(\frac{d}{dx}\right)^n e^{-\hat{c}\left(\frac{x}{2}\right)^2} \varphi_0 = H_n\left(-\hat{c}\,\frac{x}{2}, -\frac{\hat{c}}{4}\right) e^{-\hat{c}\left(\frac{x}{2}\right)^2} \varphi_0,$$

$$\tag{1.2.23}$$

whose consequences will be discussed later in this book.

1.2.2 $\hat{b}$-*Operator*

To make the previous remark even more effective, it is important to stress that not only is the exponential formal series useful, but other series can also be exploited. Introducing, for example, with the same procedure of Section 1.1, the operator $\hat{b}$, with the corresponding vacuum Φ, we provide the following statements.

Definition 3. $\forall r \in \mathbb{R}$

$$\hat{b}^r \Phi_0 := \Phi_r = \frac{1}{(\Gamma(r+1))^2}. \tag{1.2.24}$$

Lemma 4. *By using an ordinary MacLaurin expansion, $\forall x \in \mathbb{R}$, we get*

$$J_0(x) = \frac{1}{1 + \hat{b}\left(\frac{x}{2}\right)^2} \Phi_0$$

$$= \sum_{r=0}^{\infty} (-1)^r \left(\frac{x}{2}\right)^{2r} \left(\hat{b}^r \Phi_0\right) = \sum_{r=0}^{\infty} \frac{(-1)^r}{(r!)^2} \left(\frac{x}{2}\right)^{2r}, \tag{1.2.25}$$

thus expressing Bessel function in another way, according to operator[g] $\hat{b}$.

The use of the integral formula

$$\int_{-\infty}^{\infty} \frac{1}{1 + a\,x^2} dx = \frac{\pi}{\sqrt{a}}, \quad \forall a \in \mathbb{R}, \tag{1.2.26}$$

yields, for example, the integral in Corollary 2.

Corollary 2. *By applying the operator* (1.2.24),

$$\int_{-\infty}^{\infty} J_0(x)dx = 2\frac{\pi}{\sqrt{\hat{b}}} \Phi_0 = 2\,\pi\,\hat{b}^{-\frac{1}{2}} \Phi_0 = 2\frac{\pi}{(\Gamma\left(\frac{1}{2}\right))^2} = 2. \tag{1.2.27}$$

---

[g]The derivation of Eq. (1.2.25) contains a formal abuse. The series expansion of the rational umbral image should have a limited convergence radius. This is not true if we include in the analysis the properties of the $\hat{b}$ operator, which justifies the convergence of the r.h.s. $\forall x \in \mathbb{R}$.

The use of a rational function instead of an exponential as umbral image of a Bessel is therefore equally useful.

A conclusion to be drawn from the previous examples is that, **choosing an elementary function as the umbral image of another, the properties of the first can be "transferred" to the second**, provided that a set of formal rules be applied. Such a statement is referred to as[h] **Principle of Permanence of the Formal Properties**.

1.2.3   *Principle of Permanence of the Formal Properties*

The "principle" we have mentioned, a close consequence of the *Ramnujian Master Theorem* (*RMT*) (see Section 7.6) and of the umbral formalism, can also be worded as follows.

Proposition 2. *If an umbral correspondence between two different functions is established, such a correspondence can be extended to other operations including integrals.*

We illustrate such a statement by means of the following Example.

Example 7. Suppose we want to calculate the integral

$$I_T(a, b) = \int_{-\infty}^{\infty} C_0(bx)e^{-ax^2}dx, \quad \forall b \in \mathbb{R}, \forall a \in \mathbb{R}^+, \qquad (1.2.28)$$

where the subscript T stands for Tricomi. According to the umbral definition of 0-order Tricomi–Bessel function (1.2.18) and recalling (1.2.17) $C_0(bx) = J_0(2\sqrt{bx}) = e^{-\hat{c}bx}\varphi_0$, we can write

$$I_e(a, b) = \int_{-\infty}^{\infty} e^{-\hat{c}\,b\,x}e^{-ax^2}dx\,\varphi_0, \quad \forall b \in \mathbb{R}, \forall a \in \mathbb{R}^+, \qquad (1.2.29)$$

---

[h]The "principle" we have mentioned traces back to the works of George Peacock [140], referred to as analogous statement, reported, as "Principle of the permanence of equivalent forms". It was formulated in his book *Symbolical Algebra* and was enunciated as: "Whatever algebraic forms are equivalent when the symbols are general in form, but specific in value, will be equivalent likewise when the symbols are general in value as well as in form." The Authors of this book have formulated their principle unaware of the Peacock's work.

the subscript e stands for exponential. Since (see (1.2.2))

$$I(a,b) = \int_{-\infty}^{\infty} e^{bx} e^{-ax^2} dx = \sqrt{\frac{\pi}{a}} e^{\frac{b^2}{4a}}, \quad \forall b \in \mathbb{R}, \forall a \in \mathbb{R}^+, \quad (1.2.30)$$

we "invoke" the previous principle and assume that the same relation holds, $\forall b \in \mathbb{R}, \forall a \in \mathbb{R}^+$, under the correspondence

$$I_T(a,b) = I_e(a,b) = I(a,-b\hat{c})\varphi_0 = \sqrt{\frac{\pi}{a}} W_0^{(2)}\left(\frac{b^2}{4a}\right). \quad (1.2.31)$$

Before concluding this introductory chapter, we provide an application of the techniques we have developed so far to the study of Mittag–Leffler (ML) function, which is a useful "case study" for our purposes and for the applications we will mention in the Theory of Fractional Differintegrals.

1.3 Mittag–Leffler Function: An Umbral Point of View

In this section, we explore the consequence of the umbral restyling of the *ML* function [124]– [173] (Fig. 1.2)

$$E_{\alpha,\beta}(x) = \sum_{r=0}^{\infty} \frac{x^r}{\Gamma(\alpha r + \beta)}, \quad \forall x \in \mathbb{R}, \forall \alpha, \beta \in \mathbb{R}^+, \quad (1.3.1)$$

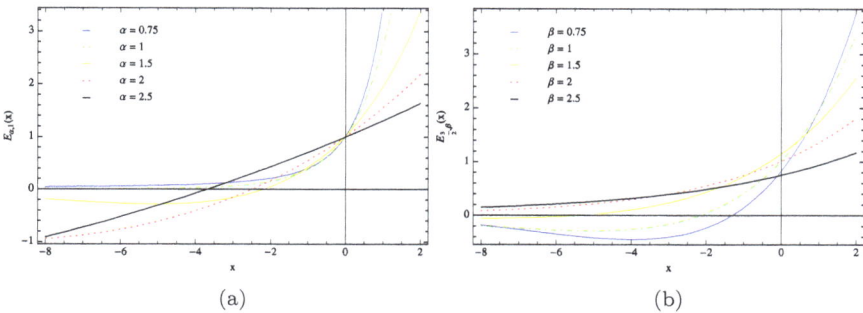

Figure 1.2: ML Functions $E_{\alpha,\beta}(x)$ for different α and β values. (a) $\beta = 1$ and different α values, (b) $\alpha = \frac{3}{2}$ and different β values.

which has become a pivotal tool of *fractional calculus* [134], namely of the branch of calculus employing derivatives or integrals of fractional order as further discussed later in this book.

According to the assumptions of the previous sections, we can cast the *ML* function in the following form [66].

Proposition 3. $\forall x \in \mathbb{R}, \forall \alpha, \beta \in \mathbb{R}^+$

$$E_{\alpha,\beta}(x) = \hat{c}^{\beta-1}\frac{1}{1 - \hat{c}^\alpha x}\varphi_0. \tag{1.3.2}$$

Proof. We have $\forall x \in \mathbb{R}, \forall \alpha, \beta \in \mathbb{R}^+$ (see (1.1.6)),

$$E_{\alpha,\beta}(x) = \sum_{r=0}^{\infty} \frac{x^r}{\Gamma(\alpha r + \beta)} \sum_{r=0}^{\infty} \hat{c}^{\alpha r+\beta-1} x^r \varphi_0 = \hat{c}^{\beta-1} \sum_{r=0}^{\infty} (\hat{c}^\alpha x)^r \varphi_0$$

$$= \hat{c}^{\beta-1}\frac{1}{1 - \hat{c}^\alpha x}\varphi_0. \tag{1.3.3}$$

$\square$

We have formally reduced the trascendental function (1.3.1) to a rational form.

In deriving the previous results, we have not paid any attention to the radius of convergence of the series in (1.3.3) since the expansion holds only in a *formal sense*, being an operator expansion. The convergence must be checked for the final function obtained via the action of the umbral operator on the vacuum and will be defined in terms of the variable x only.

By the same procedure, namely by treating $\hat{c}$ as an ordinary constant, we can recast the *ML* function in terms of an integral representation [66] through the following procedure.

Corollary 3. $\forall x \in \mathbb{R}, \forall \alpha, \beta \in \mathbb{R}^+$, *by the use of the* **Laplace transform** *identity*

$$\frac{1}{A} = \int_0^\infty e^{-sA} ds, \tag{1.3.4}$$

which holds independent of the nature of A (be it a number or an operator), we get

$$E_{\alpha,\beta}(x) = \hat{c}^{\beta-1}\frac{1}{1 - \hat{c}^\alpha x}\varphi_0 = \hat{c}^{\beta-1} \int_0^\infty e^{-s} e^{\hat{c}^\alpha x\, s}\, ds\, \varphi_0. \tag{1.3.5}$$

Corollary 4. *According to Eq. (1.1.13) and to the previous discussion, we recognize that* $\forall x \in \mathbb{R}, \forall \alpha, \beta \in \mathbb{R}^+$

$$\hat{c}^{\,\beta-1} e^{\hat{c}^{\alpha}\, x} \varphi_0 = W^{\alpha}_{\beta-1}(x) = \sum_{r=0}^{\infty} \frac{x^r}{r!\,\Gamma(\alpha\, r + \beta)}, \tag{1.3.6}$$

therefore, we end up with (*see* (1.3.5))

$$E_{\alpha,\beta}(x) = \int_0^{\infty} e^{-s} W^{\alpha}_{\beta-1}(xs)\,ds, \tag{1.3.7}$$

which states that the ML is the Borel transform of the BW function (*see Eq. (1.1.18)*) [66].

In order to provide a further flavor of the flexibility of the method we are proposing, we consider the problem of evaluating the following integral [66].

Example 8. $\forall \alpha, \beta \in \mathbb{R}^+$

$$I_{\alpha,\,\beta} = \int_{-\infty}^{\infty} E_{\alpha,\beta}(-x^2)\,dx, \tag{1.3.8}$$

which can be easily computed provided that, in the integration process, we treat as ordinary constants the operators appearing in it. We find therefore, using Eq. (1.3.2),

$$I_{\alpha,\,\beta} = \hat{c}^{\,\beta-1} \int_{-\infty}^{\infty} \frac{1}{1 + \hat{c}^{\,\alpha} x^2}\,dx\; \varphi_0, \tag{1.3.9}$$

and, exploiting the integral result (1.2.26) and the rule (1.1.6), we obtain

$$\text{(i)}\; I_{\alpha,\,\beta} = \hat{c}^{\,\beta-1} \frac{\pi}{\sqrt{\hat{c}^{\,\alpha}}} \varphi_0 = \pi\, \hat{c}^{\,\beta-\frac{\alpha}{2}-1} \varphi_0 = \frac{\pi}{\Gamma\left(\beta - \frac{\alpha}{2}\right)} \tag{1.3.10}$$

or, by using the integral representation in terms of *BW* function (see Eq. (7.8.17)) we end up with the same result, namely

$$\text{(ii)}\; I_{\alpha,\beta} = \left(\sqrt{\pi} \hat{c}^{\,\beta-\frac{\alpha}{2}-1} \int_0^{\infty} e^{-s} s^{-\frac{1}{2}}\,ds \right) \varphi_0 = \sqrt{\pi}\, \Gamma\left(\frac{1}{2}\right) \hat{c}^{\,\beta-\frac{\alpha}{2}-1} \varphi_0$$

$$= \frac{\pi}{\Gamma\left(\beta - \frac{\alpha}{2}\right)}. \tag{1.3.11}$$

The *exponential umbral image* of the *ML*, provided in Proposition 4, can be realized by the use of the following umbral operator.

Definition 4. We introduce, $\forall \alpha, \beta \in \mathbb{R}^+$, the umbral vacuum

$$\psi_\kappa := \frac{\Gamma(\kappa + 1)}{\Gamma(\alpha \kappa + \beta)}, \quad \forall \kappa \in \mathbb{R}. \tag{1.3.12}$$

Definition 5. We define the shift operator $_{\alpha,\beta}\hat{d}$, $\forall \alpha, \beta \in \mathbb{R}^+$, such that, $\forall \kappa \in \mathbb{R}$, by using the same procedure of Theorem 1, we get

$$_{\alpha,\beta}\hat{d}^\kappa \psi_0 = \frac{\Gamma(\kappa + 1)}{\Gamma(\alpha \kappa + \beta)}. \tag{1.3.13}$$

Proposition 4. $\forall \alpha, \beta \in \mathbb{R}^+, \forall x \in \mathbb{R}$, *the exponential umbral image of the ML function can be realized by*

$$E_{\alpha,\beta}(x) = e^{\,_{\alpha,\beta}\hat{d}\,x} \psi_0. \tag{1.3.14}$$

Proof. $\forall \alpha, \beta \in \mathbb{R}^+, \forall x \in \mathbb{R}$, using known results of geometrical series, the operator definition (1.3.13) and the Γ-function property (1.0.3), we obtain

$$e^{\,_{\alpha,\beta}\hat{d}x} \psi_0 = \sum_{r=0}^{\infty} \frac{x^r \left(_{\alpha,\beta}\hat{d}\right)^r}{r!} \psi_0 = \sum_{r=0}^{\infty} \frac{x^r}{\Gamma(\alpha r + \beta)} = E_{\alpha,\beta}(x). \tag{1.3.15}$$
$\square$

Now, we can obtain the same previous result ((1.3.10)–(1.3.11)) exploiting $_{\alpha,\beta}\hat{d}$-operator like in the following.

Example 9. It is enough to use Eqs. (1.3.14)-(1.2.2)-(1.3.13)-(1.0.4) to get $\forall \alpha, \beta \in \mathbb{R}^+$

$$I_{\alpha,\beta} = \int_{-\infty}^{\infty} E_{\alpha,\beta}(-x^2)dx = \int_{-\infty}^{\infty} e^{-_{\alpha,\beta}\hat{d}x^2} dx \,\psi_0 = \sqrt{\pi} \left(_{\alpha,\beta}\hat{d}\right)^{-\frac{1}{2}} \psi_0$$

$$= \frac{\pi}{\Gamma\left(\beta - \frac{\alpha}{2}\right)}. \tag{1.3.16}$$

Then, as already noted, there is no reason to privilege the exponential or the rational image function, which are easily shown to be equivalent for the derivation of results of practical interest.

We have so far outlined the essential elements of the "calculus" we will employ in the following. We will see in the subsequent chapters how the method specifically works to define, within the same formal context, special polynomials and functions. An account of the umbral treatment of Hermite, Laguerre, Legendre and Jacobi polynomials is given in Chapters 2 and 3.

Chapter 2

Umbral Formulation of Hermite Polynomials

2.1 Theory of Special Polynomials and "Monomiality" Principle

In the previous chapter, we have fixed the essential tools of Umbral calculus or, better, of our computational recipes mixing operators (of umbral or differential nature) and special transforms. We have stated "general principles" concerning the criteria to define the umbral images and the procedures to take advantage of these definitions and make progress on the properties of the associated Special functions. We have seen how the distinctions between, e.g., Bessel, Gaussian and Lorentzian functions is just a matter of defining convenient arrangements of the umbral operator, associated with the series definition of the function itself.

In this and in the following chapters, we develop an umbral point of view and realize the program of reducing the Hermite and Laguerre polynomials to a straightforward form of Newton binomial. The concepts and the formalism we develop here will be the backbone of the studies outlined in the forthcoming chapters, regarding special functions and the relevant applications.

In the first part of the chapter we present the theoretical foundations of the monomiality, a tool to treat the special polynomials as ordinary monomials

The principle of *Quasi-Monomiality* (in short Quasi-Monomials) (*QMs*) has been formulated, in its modern conception, in a series of papers listed in [24, 47–49]. Earlier suggestions trace back to the work of J. F. Steffensen [162] who introduced the concept of poweroid.

However, even earlier works, deepening their roots in the formalism associated with the calculus of differences, suggested analogous seminal ideas We will initially use the methods of monomiality for Hermite polynomials only. Later, at the end of this chapter we will extend the formalism to other family of polynomials.

Even though we will treat the problem in more general terms later in the chapter, we give here the key elements to deal with the concepts associated with the monomiality principle.

Definition 6. A family of polynomials $p_n(x)$, $\forall n \in \mathbb{N}$, $\forall x \in \mathbb{R}$, is said to be a QM if a couple of operators, $\hat{M}$ and $\hat{P}$, hereafter called *Multiplicative* and *Derivative* operators, respectively, do exist and act according to the rules

$$\hat{M} p_n(x) = p_{n+1}(x), \tag{2.1.1}$$

$$\hat{P} p_n(x) = n\, p_{n-1}(x). \tag{2.1.2}$$

The previous identities clarify the role and, hence, the names of the two operators and can be exploited to derive the relevant properties.

Corollary 5. *The combinations of the two identities yields*

$$\hat{M}\hat{P} p_n(x) = np_n(x), \quad \forall n \in \mathbb{N}, \ \forall x \in \mathbb{R} \tag{2.1.3}$$

and

$$\hat{P}\hat{M} p_n(x) = (n+1)p_n(x), \tag{2.1.4}$$

which eventually allows the conclusion that the commutator of the multiplicative and derivative operators is

$$[\hat{P}, \hat{M}] = \hat{P}\hat{M} - \hat{M}\hat{P} = \hat{1}. \tag{2.1.5}$$

It is therefore evident that $\hat{P}, \hat{M}, \hat{1}$ can be viewed as the generators of a Weyl algebra [18]. Furthermore, for a specific differential realization of the aforementioned operators, Eqs. (2.1.3)–(2.1.4) provide the eigenvalue equation of the $p_n(x)$ polynomials.

Proposition 5. *Assuming that the "vacuum" is such that $p_0(x) = 1$, we infer, from Eq. (2.1.1), that, $\forall n \in \mathbb{N}$, $\forall x \in \mathbb{R}$, we can generate* *our family of polynomials $p_n(x)$ according to the "rule"*

$$\hat{M}^n 1 = p_n(x). \tag{2.1.6}$$

Corollary 6. *The relevant generating function straightforwardly follows from Eq. (2.1.6), namely*

$$\sum_{n=0}^{\infty} \frac{t^n}{n!} p_n(x) = \sum_{n=0}^{\infty} \frac{t^n \hat{M}^n}{n!} 1 = e^{t\hat{M}} 1. \tag{2.1.7}$$

It is worth adding the further operational identity, which is a fairly direct consequence of the previous equation

$$\sum_{n=0}^{\infty} \frac{t^n}{n!} p_{n+l}(x) = \hat{M}^l e^{t\hat{M}} 1 = \partial_t^l e^{t\hat{M}} 1. \tag{2.1.8}$$

The previous remarks are a nut-shell content of the QM theory of special polynomials. The formalism will be further elaborated in the forthcoming part of the chapter, but in the remainder of this section we see how Hermite polynomials can be framed within this context and leave the forthcoming parts of the book for other "popular" families.

Example 10. In (1.2.5), we have introduced the two-variable Hermite polynomials (HP) $H_n(x, y)$, which are shown to satisfy the recurrences [18]

$$(x + 2y\partial_x)H_n(x, y) = H_{n+1}(x, y), \quad \partial_x H_n(x, y) = nH_{n-1}(x, y), \tag{2.1.9}$$

$\forall x, y \in \mathbb{R}, \forall n \in \mathbb{N}$. It is accordingly evident that Hermite polynomials are QM and that

$$\hat{M} = x + 2y\partial_x, \tag{2.1.10}$$

$$\hat{P} = \partial_x. \tag{2.1.11}$$

The second-order differential equation satisfied by $H_n(x, y)$ can therefore be written in terms of the relevant product, namely

$$\hat{M}\hat{P} = (x + 2y\partial_x) \partial_x, \tag{2.1.12}$$

yielding

$$x\partial_x H_n(x,y) + 2y\partial_x^2 H_n(x,y) = nH_n(x,y). \tag{2.1.13}$$

Regarding the generating function, by using Eqs. (2.1.7)–(2.1.10), we find

$$\sum_{n=0}^{\infty} \frac{t^n}{n!} H_n(x,y) = e^{t(x+2y\partial_x)} 1. \tag{2.1.14}$$

The use of the Weyl rule (see Eq. (25) in the last chapter) [77, 170] yields[a]

$$e^{t(x+2y\partial_x)} 1 = e^{-\frac{1}{2}[tx,2yt\partial_x]} e^{tx} e^{2yt\partial_x} 1 = e^{xt+yt^2}, \tag{2.1.15}$$

thus providing the quoted generating function (1.2.6).

In Eq. (2.1.15), the last exponential operator disappears because

$$e^{2yt\partial_x} 1 = 1. \tag{2.1.16}$$

We may however ask how the previous operator identities should be modified if, instead of a constant, it acts on a generic function of the variable x. We have noted (2.1.6)–(2.1.10) that

$$(x + 2y\partial_x)^n 1 = H_n(x,y), \tag{2.1.17}$$

but if we relax the assumption that the operator on the left is acting on unity, the result is however slightly more complicated.

We first note that (series expansions (1.2.6))

$$\sum_{n=0}^{\infty} \frac{t^n}{n!}(x + 2y\partial_x)^n = e^{yt^2+xt} e^{2ty\partial_x} = \sum_{m=0}^{\infty} \frac{t^m}{m!} H_m(x,y) \sum_{s=0}^{\infty} \frac{(2ty)^s}{s!} \partial_x^s$$

$$= \sum_{n=0}^{\infty} \frac{t^n}{n!} \hat{N}_n,$$

$$\hat{N}_n = \sum_{s=0}^{n} \binom{n}{s} (2y)^s H_{n-s}(x,y) \partial_x^s. \tag{2.1.18}$$

---

[a]We remind that $[x, \partial_x] = -1$.

Equating the same like power t terms, we end up with the so-called *Burchnall* identity [77]

$$(x + 2y\partial_x)^n = \sum_{s=0}^{n} \binom{n}{s} (2y)^s H_{n-s}(x, y)\partial_x^s, \qquad (2.1.19)$$

useful to evaluate the action of the $\hat{M}$-Hermite operator on a generic function, namely

$$(x + 2y\partial_x)^n f(x) = \sum_{s=0}^{n} \binom{n}{s} (2y)^s H_{n-s}(x, y) f^{(s)}(x), \qquad (2.1.20)$$

where $f^{(s)}(x)$ denotes the sth-derivative of the function $f(x)$ and, therefore, Eq. (2.1.20) reduces to Eq. (2.1.17) if $f(x)$ is a constant.

Example 11. The identity (2.1.8) applied to the case of HP, through the result in Eq. (2.1.15), yields

$$\sum_{n=0}^{\infty} \frac{t^n}{n!} H_{n+l}(x, y) = \partial_t^l\, e^{xt+yt^2}. \qquad (2.1.21)$$

The successive derivative with respect to t can be evaluated using the following procedure[b]:

$$\sum_{l=0}^{\infty} \frac{u^l}{l!}\partial_t^l\, e^{xt+yt^2} = e^{u\partial_t}\, e^{xt+yt^2} = e^{x(t+u)+y(t+u)^2}$$

$$= e^{ux+2yut+u^2 y}e^{xt+yt^2}$$

$$= \sum_{m=0}^{\infty} \frac{u^m}{m!} H_m(x + 2yt, y)e^{xt+yt^2}. \qquad (2.1.22)$$

Therefore, equating the u-like power coefficient, we find

$$\partial_t^l\, e^{xt+yt^2}\, H_l(x + 2yt, y)e^{xt+yt^2}, \qquad (2.1.23)$$

thus finally getting

$$\sum_{n=0}^{\infty} \frac{t^n}{n!} H_{n+l}(x, y) = \partial_t^l\, e^{xt+yt^2} = H_l(x + 2yt, y)e^{xt+yt^2}. \qquad (2.1.24)$$

---

[b]We used the shift operator [18] many times in this book, see, e.g., Chapters 1 or 7, Eq. (7.3.1).

Remark 2. In Proposition 8, Section 2.3, we will see how the Hermite polynomials can be represented by the Newton binomial umbral image $(x + {}_y\hat{h})^n\vartheta_0$. Here we indicate the Hermite operator with

$$\hat{h}_{x,y}\theta_0 = (x + {}_y\hat{h})\theta_0 \qquad (2.1.25)$$

and write the generating function in Eq. (2.1.24) according to $\hat{h}_{x,y}$ umbral operator and using the HP generating function (1.2.6)

$$G_l(x,y;t) = \sum_{n=0}^{\infty} \frac{t^n}{n!} H_{n+l}(x,y) = \hat{h}_{x,y}^l \sum_{n=0}^{\infty} \frac{t^n}{n!} \hat{h}_{x,y}^n \theta_0 = \hat{h}_{x,y}^l e^{t\,\hat{h}_{x,y}}\theta_0$$

$$= \partial_t^l e^{t\,\hat{h}_{x,y}}\theta_0 = \partial_t^l e^{xt+yt^2} = H_l(x+2yt,t)e^{xt+yt^2}.$$

$$(2.1.26)$$

We have mentioned the monomiality of the Hermite family, in the forthcoming section we will recognize that of Laguerre polynomials and, in the chapter devoted to complements and exercises, those of Appéll and Sheffer sequences.

In the forthcoming sections, we provide a description of the orthogonal properties of HP within the so far developed operational framework.

2.2   Hermite Polynomial, Orthogonal Properties and Operational Formalism

In this section, we discuss the orthogonal nature of the two-variable HP using the rules we have established in the previous section. The point of view we describe is slightly different from the conventional treatment and is developed to obtain a more general definition of the concepts, underlying the orthogonality properties of different polynomial families, we will employ in the course of this book.

Lemma 5. *Along with the definition of QM, we can also introduce functions expanded on QMs, according to the rule (we omit the vacuum 1, for conciseness)*

$$f(\hat{M}) = \sum_{n=0}^{\infty} a_n \hat{M}^n = \sum_{n=0}^{\infty} a_n p_n(x), \qquad (2.2.1)$$

which can be understood as an expansion over the QM basis $p_n(x), \forall x \in \mathbb{R}.$

The meaning of Eq. (2.2.1) can be worded as follows.

Proposition 6. *Any function having a MacLaurin expansion on the ordinary monomial basis can be associated with a corresponding function expanded on its QM counterpart.*

Therefore, once we have established a correspondence between $f(\hat{M})$ and a function of x, we have found the expansion of that function on the QM basis. We provide a first result in terms of HP.

Example 12. $\forall x, y \in \mathbb{R}$ (see [18] and (2.1.17))

$$e^{y\partial_x^2} e^x = \sum_{n=0}^{\infty} \frac{e^{y\partial_x^2} x^n}{n!} = \sum_{n=0}^{\infty} \frac{(x + 2y\partial_x)^n}{n!} = \sum_{n=0}^{\infty} \frac{H_n(x, y)}{n!} = e^{x+y},$$

$$(2.2.2)$$

or (see Eq. (2.2.15) for further details)

$$e^{y\partial_x^2} e^{-x^2} = \sum_{n=0}^{\infty} \frac{(-1)^n e^{y\partial_x^2} x^{2n}}{n!}$$

$$= \sum_{n=0}^{\infty} \frac{(-1)^n}{n!} H_{2n}(x, y) = \frac{1}{\sqrt{1 + 4y}} e^{-\frac{x^2}{1+4y}}. \quad (2.2.3)$$

The second series has a limited convergence radius $\mid y \mid < \frac{1}{4}$ as further discussed later.

It is evident that the previous statement does not allow us to conclude that having recognized the monomial nature of polynomials is not sufficient to state the relevant horthogonality. Within this respect, a more elaborated treatment, reported in what follows, is necessary. We settle out the problem starting from the following example.

Example 13. We assume that a function $f(x)$ can be expanded as

$$f(x) = \sum_{n=0}^{\infty} a_n H_n(x, y), \qquad (2.2.4)$$

where a_n are the coefficients of the expansion. It is evident that, since f is a function of one variable only, y should be regarded as a parameter. Let us note that, since $H_n(x, y) = e^{y\partial_x^2} x^n$ [18], we find, from Eqs. (2.2.1)–(2.2.2),

$$e^{-y\partial_x^2} f(x) = \sum_{n=0}^{\infty} a_n x^n. \tag{2.2.5}$$

To ensure the existence of a Gauss–Weierstrass transform (1.2.2) of $f(x)$, we assume that $\mathbf{y}$ is a *negative* defined parameter by setting $y = -\mid y \mid, \forall y \in \mathbb{R}$, thus, writing Eq. (2.2.5) (by applying (1.2.2) at Eq. (2.2.5) and through a variable change) as

$$\sum_{n=0}^{\infty} a_n x^n = \frac{1}{2\sqrt{\pi \mid y \mid}} \int_{-\infty}^{\infty} e^{-\frac{(x-\xi)^2}{4\mid y \mid}} f(\xi) d\xi. \tag{2.2.6}$$

It is therefore evident that

$$\sum_{n=0}^{\infty} a_n x^n = \frac{1}{2\sqrt{\pi \mid y \mid}} \int_{-\infty}^{\infty} e^{\frac{x\xi}{2\mid y \mid} - \frac{x^2}{4\mid y \mid}} \left(e^{-\frac{\xi^2}{4\mid y \mid}} f(\xi) \right) d\xi$$

$$= \frac{1}{2\sqrt{\pi \mid y \mid}} \sum_{m=0}^{\infty} \frac{x^m}{m!} \int_{-\infty}^{\infty} H_m \left(\frac{\xi}{2 \mid y \mid}, -\frac{1}{4 \mid y \mid} \right) \left(e^{-\frac{\xi^2}{4\mid y \mid}} f(\xi) \right) d\xi$$

$$\tag{2.2.7}$$

and, by equating x-like power, we obtain for the expansion coefficients

$$a_m = \frac{1}{2m!\sqrt{\pi \mid y \mid}} \int_{-\infty}^{\infty} H_m \left(\frac{\xi}{2 \mid y \mid}, -\frac{1}{4 \mid y \mid} \right) e^{-\frac{\xi^2}{4\mid y \mid}} f(\xi) d\xi. \tag{2.2.8}$$

The function

$$u_m \left(x, - \mid y \mid \right) = \frac{1}{2m!\sqrt{\pi \mid y \mid}} H_m \left(\frac{x}{2 \mid y \mid}, -\frac{1}{4 \mid y \mid} \right) e^{-\frac{x^2}{4\mid y \mid}} \tag{2.2.9}$$

is bi-orthogonal to $H_m \left(x, - \mid y \mid \right)$, namely

$$\int_{-\infty}^{\infty} u_m \left(x, - \mid y \mid \right) H_n \left(x, - \mid y \mid \right) dx = \delta_{m,n}. \tag{2.2.10}$$

We note that, within the present context, $\mid y \mid$ is only a parameter and that, setting, e.g., $\mid y \mid = \frac{1}{2}$, we obtain the ordinary Hermite functions of quantum harmonic oscillator [23, 57].

It is worth noting that, to ensure the condition (2.2.10), it is necessary that the Re $(\mid y \mid)$ be positive.[c]

Even though trivial, we provide a further example of expansion associated with the *shifted HP* defined in the following.

Example 14. Let the operational rule (either y and z are parameters) [18]

$$H_n(x, y; z) := H_n(x - z, y) = e^{-z\partial_x + y\partial_x^2} x^n. \qquad (2.2.11)$$

The conditions for the expansion on this family of polynomials occurs through the same procedure as before, we set therefore (the conditions on the sign of z are not influent)

$$f(x) = \sum_{n=0}^{\infty} a_n H_n (x, - \mid y \mid; z) \qquad (2.2.12)$$

and therefore

$$e^{\mid y \mid \partial_x^2 + z\partial_x} f(x) = \sum_{n=0}^{\infty} a_n x^n. \qquad (2.2.13)$$

Using the same procedure of Example 13 we end up with the following family of bi-orthogonal functions to the shifted HP

$$u_n (x, - \mid y \mid; z) = \frac{1}{2m!\sqrt{\pi \mid y \mid}} H_m \left(\frac{x - z}{2 \mid y \mid}, -\frac{1}{4 \mid y \mid} \right) e^{-\frac{(x-z)^2}{4\mid y \mid}}, \qquad (2.2.14)$$

which provides the Hermite functions of the shifted harmonic oscillator.

The expansion (2.2.3), on the light of the previous discussion, can be explained as follows.

---

[c]It is therefore clear that, in order that $u_m(.,.)$ and $H_n(.,.)$ are orthogonal to each other, it is necessary that the parameter y has a negative real part.

Proof. We prove Eq. (2.2.3). By expanding e^{-x^2}, we get

$$e^{y\partial_x^2}e^{-x^2} = e^{y\partial_x^2}\sum_{r=0}^{\infty}\frac{(-1)^r}{r!}x^{2r}, \qquad (2.2.15)$$

interchanging the summation index and the exponential operator, we get

$$e^{y\partial_x^2}e^{-x^2} = \sum_{r=0}^{\infty}\frac{(-1)^r}{r!}e^{y\partial_x^2}x^{2r} = \sum_{r=0}^{\infty}\frac{(-1)^r}{r!}H_{2r}(x,y). \qquad (2.2.16)$$

From the other side, by using the GWI (1.2.2) we easily infer the Glaisher identity (7.8.56) $e^{y\partial_x^2}e^{-x^2} = \dfrac{1}{\sqrt{1+4y}}e^{-\frac{x^2}{1+4y}}$, which justifies what is reported in Eq. (2.2.3). It is important to stress that, while the transform integral of $e^{y\partial_x^2}e^{-x^2}$ converges for any $y > -\frac{1}{4}$, the expansion in terms of two-variable Hermite has a limited range of convergence (see later (2.3.10)). The reasons rely on the fact that the interchanging of the summation and the exponential operator is admitted in the previously quoted interval of convergence.   □

Remark 3. Before closing this section let us note that the expansion of a monomial in terms of HP is fairly straightforward and is expressed as

$$x^n = \sum_{r=0}^{\lfloor\frac{n}{2}\rfloor} a_{n,r}H_{n-2r}(x,y), \qquad (2.2.17)$$

which yields the condition

$$a_{n,r} = \frac{n!}{(n-2r)!r!}(-y)^r, \qquad (2.2.18)$$

which, being a finite sum, does require the condition $y = -\mid y \mid$.

We can further stress the role of the y parameter in the theory of two-variable Hermite by noting that

(a) Being $H_n(x,y)$ solutions of the heat equation [178], they play the role of the so-called *heat polynomials*, introduced by *Widder* in [178], where y is associated with time;

(b) A more subtle meaning can be inferred from the plots reported in Fig. 2.1 in which we have shown the polynomials in an x, y, z space. Within this context, the operational rule $H_n(x, y) = e^{y\partial_x^2} x^n$ [18] can be understood in a geometrical sense. *The exponential operator transforms an ordinary monomial into a Hermite type special polynomial.* The "evolution" from an ordinary monomial to the corresponding Hermite is shown by moving the cutting plane orthogonal to the y axis. For a specific value of the polynomial degree n, the polynomials lie on the cutting plane, as shown in the figures. It is worth stressing that *only for negative values of y* do the *polynomials exhibit zeros,* in accordance with the fact that in this region they realize an orthogonal set.

The discussions and the examples we have developed so far show operational point of view on the theory of special polynomials and open many new possibilities of interpretations on the nature of the polynomials themselves as will be further discussed in the forthcoming sections of this chapter.

2.3   An Umbral Point of View on Hermite Polynomials

In this section, we consider the transition from Monomiality to Umbral interpretation of *HP*. We noted, e.g., in Eqs. (1.2.17) and (1.2.25), that the Gaussian function is the umbral image of the cylindrical Bessel function, in this paragraph we also touch on that the *Newton binomial* realizes the umbral image of *HP*. The umbral nature of *HP* is sporadically reported in the mathematical literature, the link with the present formulation will be shortly discussed in Chapter 7.

Definition 7. We introduce

$$\theta(z) := \theta_z = y^{\frac{z}{2}} \left(\frac{\Gamma(z+1)}{\Gamma\left(\frac{1}{2}z+1\right)} \left| \cos\left(\frac{\pi}{2}z\right) \right| \right), \quad \forall z \in \mathbb{R}, \qquad (2.3.1)$$

the Hermite function vacuum.

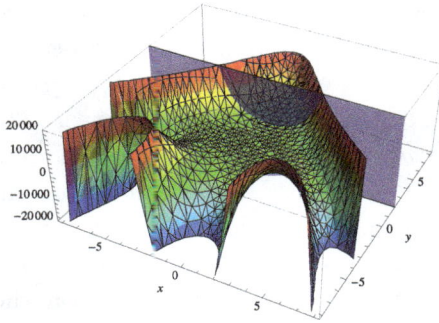

(a) $H_6(x, y)$ be cut by the plane $y = 2$.

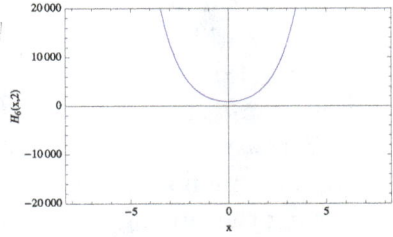

(b) $H_6(x, 2)$.

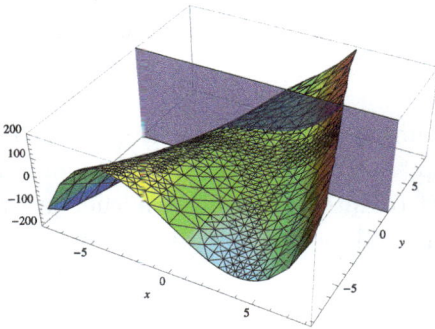

(c) $H_3(x, y)$ be cut by the plane $y = 2$.

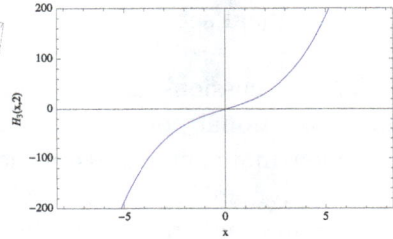

(d) $H_3(x, 2)$.

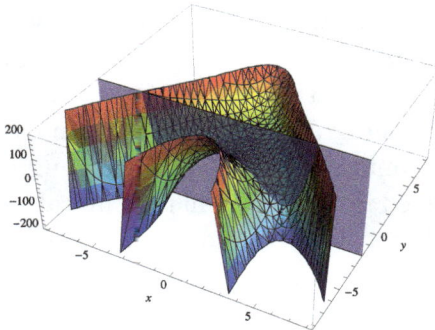

(e) $H_4(x\ y)$ be cut by the plane $y = -2$.

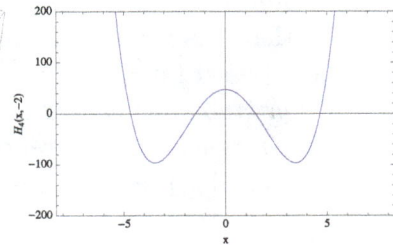

(f) $H_4(x, -2)$.

Figure 2.1:   Geometrical representation of two-variable Hermite polynomials in 3D and 2D, for different n and y values [9].

Proposition 7. *The umbral operator* $_y\hat{h}^r$ *acts on the vacuum* θ_0 *according to the rule*

$$_y\hat{h}^r\,\theta_0 := \theta_r, \quad \forall r \in \mathbb{R},$$

$$\theta_r = \frac{y^{\frac{r}{2}}r!}{\Gamma\left(\frac{r}{2}+1\right)}\left|\cos\left(r\frac{\pi}{2}\right)\right| = \begin{cases} 0 & r = 2s+1 \\ y^s\dfrac{(2s)!}{s!} & r = 2s \end{cases} \quad \forall s \in \mathbb{Z}.$$

(2.3.2)

Proof. $\forall s \in \mathbb{Z}, \forall r, y, z \in \mathbb{R}$, by using the shift operator $_y\hat{h} = e^{\partial_z}$, we get

$$_y\hat{h}^r\theta_0 = {_y\hat{h}^r}\theta(z)\Big|_{z=0} = e^{r\partial_z}\theta(z)\Big|_{z=0} = \theta(z+r)\big|_{z=0} = \theta_{z+r}\big|_{z=0}$$

$$= y^{\frac{z+r}{2}}\left(\frac{\Gamma(z+r+1)}{\Gamma\left(\frac{1}{2}(z+r)+1\right)}\left|\cos\left(\frac{\pi}{2}(z+r)\right)\right|\right)\Bigg|_{z=0}$$

$$= \frac{y^{\frac{r}{2}}r!}{\Gamma\left(\frac{r}{2}+1\right)}\left|\cos\left(r\frac{\pi}{2}\right)\right| = \begin{cases} 0 & r = 2s+1 \\ y^s\dfrac{(2s)!}{s!} & r = 2s \end{cases} = \theta_r.$$

$\square$

The numbers $\frac{(2s)!}{s!} = 1, 2, 12, 120, 1680, \ldots$ are recognized as the *quadrupal factorial numbers*, reported[d] in *OEIS* sequence A001813. We call the following Proposition already stated in [64].

Proposition 8. *The Newton binomial umbral version of Hermite polynomials is accordingly obtained from Eq. (2.3.2)*

$$H_n(x,y) = \left(x + {_y\hat{h}}\right)^n\theta_0, \quad \forall x, y \in \mathbb{R}, \forall n \in \mathbb{N}.$$

(2.3.3)

Proof. $\forall x, y \in \mathbb{R}, \forall n \in \mathbb{N}, \forall s \in \mathbb{Z}$, by the use of Newton binomial and Eq. (2.3.2)

$$\left(x + {_y\hat{h}}\right)^n\theta_0 = \sum_{r=0}^{n}\binom{n}{r}x^{n-r}\,_y\hat{h}^r\theta_0 = \sum_{r=0}^{n}\binom{n}{r}x^{n-r}y^s\frac{(2s)!}{s!}$$

$$= \sum_{s=0}^{\lfloor\frac{n}{2}\rfloor}x^{n-2s}y^s\frac{n!}{(n-2s)!s!} = H_n(x,y)$$

$\square$

---

[d]*OEIS* is for On-line Encyclopedia of Integer Sequences https://oeis.org/.

Corollary 7. *The correspondence between umbral* (2.3.3) *and mono-miality operators* (*Definition* 6) *is*

$$\hat{M} \leftrightarrow \left(x + {}_y\hat{h} \right). \qquad (2.3.4)$$

The generating function of HP is straightforwardly inferred (see also (2.1.7))

$$\sum_{n=0}^{\infty} \frac{t^n}{n!} H_n(x, y) = \sum_{n=0}^{\infty} \frac{\left(t\left(x + {}_y\hat{h} \right) \right)^n}{n!} = e^{xt} e^{{}_y\hat{h}t} \theta_0, \quad \forall t \in \mathbb{R}, \quad (2.3.5)$$

which yields the ordinary expression (1.2.6) by noting the following statement.

Observation 2.

$$e^{{}_y\hat{h}t} \theta_0 = \sum_{r=0}^{\infty} \frac{t^r}{r!} {}_y\hat{h}^r \theta_0 = e^{yt^2}, \quad \forall y, t \in \mathbb{R}. \qquad (2.3.6)$$

Proof. $\forall y, t \in \mathbb{R}, \theta_0$ the vacuum of the ${}_y\hat{h}$-operator, by using series expansion and Eq. (2.3.2), we obtain

$$e^{{}_y\hat{h}t} \theta_0 = \sum_{r=0}^{\infty} \frac{t^r}{r!} {}_y\hat{h}^r \theta_0 = \sum_{r=0}^{\infty} \frac{t^r}{r!} y^s \frac{(2s)!}{s!} = \sum_{s=0}^{\infty} \frac{t^{2s}}{s!} y^s = e^{yt^2}. \qquad (2.3.7)$$

$\square$

The umbral point of view to HP is particularly useful for a straightforward derivation of the relevant properties. We provide some examples through which we establish some of these properties and give an idea of the implication offered by the present formalism. We start noting the following HP properties.

Corollary 8. *Regarding the derivation of generating function involving even index HP, the following identity holds* (*applying* (2.3.3)):

$$\sum_{n=0}^{\infty} \frac{t^n}{n!} H_{2n}(x, y) = e^{t\left(x + {}_y\hat{h} \right)^2} \theta_0. \qquad (2.3.8)$$

Furthermore, on account of the GII (1.2.3) and Eq. (2.3.6), we can write

$$e^{t(x+y\hat{h})^2}\theta_0 = \frac{1}{\sqrt{\pi}}\int_{-\infty}^{\infty} e^{-\xi^2+2\sqrt{t}(x+y\hat{h})\xi}\, d\xi\, \theta_0$$

$$= \frac{1}{\sqrt{\pi}}\int_{-\infty}^{\infty} e^{-\xi^2} e^{2\sqrt{t}x\xi}\left(e^{2\sqrt{t}\xi y\hat{h}}\theta_0\right) d\xi$$

$$= \frac{1}{\sqrt{\pi}}\int_{-\infty}^{\infty} e^{-\xi^2} e^{2\sqrt{t}x\xi} e^{4yt\xi^2}\, d\xi = \frac{1}{\sqrt{1-4yt}}e^{\frac{x^2 t}{1-4yt}},$$

$$(2.3.9)$$

thus getting

$$\sum_{n=0}^{\infty}\frac{t^n}{n!}H_{2n}(x,y) = \frac{1}{\sqrt{1-4yt}}e^{\frac{x^2 t}{1-4yt}}, \quad |t| < \frac{1}{|4y|}, \qquad (2.3.10)$$

which is a result analogous to that already viewed in Eq. (2.2.3), within the framework of the orthogonal properties of two-variable Hermite.

Example 15. By exploiting Eq. (2.3.6), we get

$$e^{-yx^4} = e^{-iy\hat{h}x^2}\theta_0, \quad \forall x \in \mathbb{R}, \forall y \in \mathbb{R}_0^+. \qquad (2.3.11)$$

According to this identity, the *super-Gaussian of order 4* can be treated as an ordinary Gaussian. It is, accordingly, instructive to note that

$$\int_{-\infty}^{\infty} e^{-yx^4}\, dx = \int_{-\infty}^{\infty} e^{-iy\hat{h}x^2}\, dx\,\theta_0 = \sqrt{\frac{\pi}{iy\hat{h}}}\,\theta_0 \qquad (2.3.12)$$

and, the integral being a real integral (to use "i" is an artifice), we calculate

$$\sqrt{\pi}\left|\left(i^{-\frac{1}{2}}\right)\right|_y\hat{h}^{-\frac{1}{2}}\theta_0 = \sqrt{\pi}\frac{\sqrt{2}}{2}\frac{y^{-\frac{1}{4}}\Gamma\left(\frac{1}{2}\right)}{\Gamma\left(\frac{3}{4}\right)} = \frac{1}{2\sqrt[4]{y}}\Gamma\left(\frac{1}{4}\right), \quad (2.3.13)$$

obtaining, therefore, the correct (well-known) result integral of the super-Gaussian in Eq. (2.3.11).

Example 16. Let

$$I(\alpha, \beta) = \int_{-\infty}^{\infty} e^{-\alpha x^2 - \beta x^4} dx, \quad \forall \alpha \in \mathbb{R}, \ \forall \beta \in \mathbb{R}_0^+, \qquad (2.3.14)$$

which can be written as

$$I(\alpha, \beta) = \int_{-\infty}^{\infty} e^{-x^2(\alpha + {}_\beta \hat{h})} dx \, \theta_0, \quad \hat{H}(\alpha, \beta) = \left(\alpha + {}_\beta \hat{h}\right), \quad (2.3.15)$$

which yields

$$I(\alpha, \beta) = \sqrt{\frac{\pi}{\hat{H}(\alpha, \beta)}} \theta_0 = \sqrt{\pi} \hat{H}_{-\frac{1}{2}}(\alpha, \beta) \theta_0. \qquad (2.3.16)$$

The previous identity suggests the possibility of defining negative-order *HP*, in which the index is not constrained to integers but may keep any real value, so we give the following definition.

Definition 8. According to Eq. (2.3.3), $\forall \nu \in \mathbb{R}^+$, $\forall x, y \in \mathbb{R}$, we introduce the *Negative-Order Hermite* (NOH)

$$H_{-\nu}(x, y) = (x + {}_y \hat{h})^{-\nu} \theta_0. \qquad (2.3.17)$$

They are no more polynomials but *Hermite functions.*

Proposition 9. *The relevant NOH-function integral representation can be written as*

$$H_{-\nu}(x, y) = \frac{1}{\Gamma(\nu)} \int_0^{\infty} s^{\nu - 1} e^{-sx} e^{-ys^2} ds, \quad \forall x \in \mathbb{R}, \ \forall y, \nu \in \mathbb{R}^+.$$
$$(2.3.18)$$

Proof. $\forall x \in \mathbb{R}$, $\forall y, \nu \in \mathbb{R}^+$, by the use of Laplace transform and Eq. (2.3.6), we obtain

$$H_{-\nu}(x, y) = \frac{1}{\left(x + {}_y \hat{h}\right)^{\nu}} \theta_0 = \int_0^{\infty} e^{-xs} \frac{s^{\nu - 1} e^{-y \hat{h} s}}{\Gamma(\nu)} ds \, \theta_0$$

$$= \frac{1}{\Gamma(\nu)} \int_0^{\infty} s^{\nu - 1} e^{-xs} \left(e^{-sy \hat{h}_x} \theta_0\right) ds$$

$$= \frac{1}{\Gamma(\nu)} \int_0^{\infty} s^{\nu - 1} e^{-sx} e^{-ys^2} ds. \qquad \square$$

The use of the same procedure leads to the derivation of the infinite integral $\forall y \in \mathbb{R}$

$$I_\nu(x, y \mid m) = \int_0^\infty e^{-s^m(x+ys^m)} s^{\nu-1} ds = \int_0^\infty e^{-s^m(x+-|y|\hat{h})} s^{\nu-1} ds\, \theta_0$$

$$= \frac{\Gamma\left(\frac{\nu}{m}\right)}{m} H_{-\frac{\nu}{m}}(x, y). \tag{2.3.19}$$

In these two last sections, we have presented the theory of *HP* in terms of a non standard procedure, which allows significant degrees of freedom from the computational point of view and opens new possibilities for the solution of practical problems. A flavor of the flexibility offered by the theoretical environment we have envisaged is offered by the examples reported in the forthcoming section.

2.4 Hermite Calculus

The Hermite calculus investigated so far, useful to treat computations involving Hermite polynomials and their generalizations as well, can be expanded as follows [59].

Example 17. $\forall \alpha, \beta \in \mathbb{R} : \alpha + \beta > 0$, $\forall \gamma \in \mathbb{R}$, we consider the integral

$$I(\alpha, \beta, \gamma) = \int_{-\infty}^\infty e^{-(\alpha+\beta)x^2 - \gamma x} dx, \tag{2.4.1}$$

which can be evaluated through the *GWI* (1.2.2), thus getting

$$I(\alpha, \beta, \gamma) = \sqrt{\frac{\pi}{\alpha + \beta}} e^{\frac{\gamma^2}{4(\alpha+\beta)}}. \tag{2.4.2}$$

We restyle Eq. (2.4.1) in an umbral way

$$I(\alpha, \beta, \gamma) = \int_{-\infty}^\infty e^{-\alpha x^2 - \hat{h}_{(\gamma, -\beta)} x} dx\, \eta_0, \tag{2.4.3}$$

where we have introduced the notation

$$e^{-\hat{h}_{(\gamma, -\beta)} x} \eta_0 = \sum_{r=0}^\infty \frac{(-x)^r}{r!} \hat{h}_{(\gamma, -\beta)}^r \eta_0 = \sum_{r=0}^\infty \frac{(-x)^r}{r!} H_r(\gamma, -\beta), \tag{2.4.4}$$

based on the use of the umbral identity [64, 171]

$$\hat{h}^r_{(\gamma,-\beta)}\eta_0 = \eta_r = H_r(\gamma,-\beta). \tag{2.4.5}$$

In the integral in Eq. (2.4.3), we have treated the term which can be expanded in terms of Hermite polynomials as a single block and we have enucleated the variable x raised to the first power.

According to (1.2.2), we can write [64, 157]

$$I(\alpha,\beta,\gamma) = \sqrt{\frac{\pi}{\alpha}}e^{\frac{\hat{h}^2_{(\gamma,-\beta)}}{4\alpha}}\eta_0 = \sqrt{\frac{\pi}{\alpha}}\sum_{r=0}^{\infty}\frac{1}{r!}\left(\frac{\hat{h}^2_{(\gamma,-\beta)}}{4\alpha}\right)^r\eta_0, \tag{2.4.6}$$

which provides us with the correct result for the problem we are studying. The application of the previous prescription yields,[e] indeed, if $\left|\frac{\beta}{\alpha}\right| < 1$

$$\sqrt{\frac{\pi}{\alpha}}\sum_{r=0}^{\infty}\frac{1}{r!}\left(\frac{\hat{h}^2_{(\gamma,-\beta)}}{4\alpha}\right)^r\eta_0 = \sqrt{\frac{\pi}{\alpha}}\sum_{r=0}^{\infty}\frac{1}{r!}\frac{1}{(2\sqrt{\alpha})^{2r}}H_{2r}(\gamma,-\beta)$$

$$= \sqrt{\frac{\pi}{\alpha}}\sum_{r=0}^{\infty}\frac{1}{r!}H_{2r}\left(\frac{\gamma}{2\sqrt{\alpha}},-\frac{\beta}{4\alpha}\right)$$

$$= \sqrt{\frac{\pi}{\alpha+\beta}}e^{\frac{\gamma^2}{4(\alpha+\beta)}}. \tag{2.4.7}$$

Properties 3. The umbral operator defined in Eq. (2.4.5) satisfies the identity[f] (see (1.1.7))

$$\hat{h}^m\hat{h}^r = \hat{h}^{m+r}, \tag{2.4.8}$$

$\forall m, r \in \mathbb{R}$. It is also fairly natural to set

$$\partial_{\hat{h}}\hat{h}^r\eta_0 = r\,\hat{h}^{r-1}\eta_0 = rH_{r-1}(\gamma,-\beta). \tag{2.4.9}$$

Recalling the recurrence (2.1.9) $\partial_\gamma H_r(\gamma,-\beta) = rH_{r-1}(\gamma,-\beta)$, the "derivative" operator can be identified with

$$\partial_{\hat{h}} \to \partial_\gamma. \tag{2.4.10}$$

---

[e]Equation (2.4.7) is obtained after using the identity (2.3.10). We remind the property (1.2.10) $a^n H_n(x,y) = H_n(ax, a^2 y)$ [18].
[f]The subscript $(\gamma,-\beta)$ has been omitted because the identity holds for $\hat{h}$ operators with the same basis, hereafter it will be included whenever necessary.

Furthermore, since

$$\hat{h}\hat{h}^r \eta_0 = \hat{h}^{r+1}\eta_0 = H_{r+1}(\gamma, -\beta) \tag{2.4.11}$$

and, on account of the recurrence (2.1.9),

$$H_{r+1}(\gamma, -\beta) = \gamma H_r(\gamma, -\beta) - 2\,\beta\,r\,H_{r-1}(\gamma, -\beta), \tag{2.4.12}$$

we can also conclude that $\hat{h}$ itself can be identified with the differential operator

$$\hat{h} = \gamma - 2\,\beta\,\partial_\gamma. \tag{2.4.13}$$

in according to $\hat{M}$-operator (2.1.10).

Lemma 6. $\forall x, r, \gamma, \beta \in \mathbb{R}$,

$$\partial_x^r e^{-\hat{h}x}\eta_0 = (-1)^r \hat{h}^r e^{-\hat{h}x}\eta_0 = (-1)^r \sum_{n=0}^{\infty} \frac{(-x)^n}{n!}\hat{h}^{n+r}\eta_0$$

$$= (-1)^r \sum_{n=0}^{\infty} \frac{(-x)^n}{n!} H_{n+r}(\gamma, -\beta) \tag{2.4.14}$$

and, according to the identity [18]

$$\sum_{n=0}^{\infty} \frac{t^n}{n!} H_{n+l}(x, y) = H_l(x + 2yt, y)e^{xt+yt^2}, \tag{2.4.15}$$

we can establish the "rule"

$$\hat{h}^r e^{-\hat{h}x}\eta_0 = H_r(\gamma + 2\beta x, -\beta)e^{-(\gamma x + \beta x^2)}. \tag{2.4.16}$$

It is also worth noting that we can take a step further through the Example 18.

Example 18. Let

$$I(\gamma, \beta) = \int_{-\infty}^{\infty} e^{-\hat{h}x^2}dx\,\eta_0, \quad \forall \alpha \in \mathbb{R}, \ \forall \beta \in \mathbb{R}_0^+,$$
$$e^{-\hat{h}x^2}\eta_0 = e^{-(\gamma x^2 + \beta x^4)}, \tag{2.4.17}$$

which, after applying the *GWI* (1.2.2), writes

$$I(\gamma, \beta) = \sqrt{\pi}\,\hat{h}^{-\frac{1}{2}}\eta_0, \tag{2.4.18}$$

which makes sense only if we can provide a meaning for $\hat{h}^{-\frac{1}{2}}$. The most natural conclusion is that they can be understood as *fractional-order Hermite*, which for our purposes can be defined as follows [105]:

$$H_\nu(x, -y) = y^{\frac{\nu}{2}} \frac{e^{\frac{x^2}{4y}}}{\sqrt{\pi}} \int_0^\infty e^{-\frac{t^2}{4}} t^\nu \cos\left(\frac{x}{2\sqrt{y}}t - \frac{\pi}{2}\nu\right) dt, \quad \forall \nu \in \mathbb{R},$$

(2.4.19)

or as

$$H_\nu(x, -y) = \Gamma(\nu + 1) \sum_{r=0}^\infty \frac{x^{\nu-2r}(-y)^r}{\Gamma(\nu + 1 - 2r)r!}, \qquad x \gg y, \quad (2.4.20)$$

which has a limited range of convergence. The correctness of Eq. (2.4.18) can be readily proved involving either the definitions (2.4.19) and (2.4.20) (and it confirms the result obtained in Example 16 through another operator).

We will comment in Examples 49 and 50 on the extension of the Hermite polynomials to non-integer index.

Lemma 7. *Let us now consider the following repeated derivatives*

$$\partial_x^n e^{-\hat{h}x^2} \eta_0 = (-1)^n H_n(2\,\hat{h}\,x, -\hat{h})e^{-\hat{h}x^2}\eta_0$$

$$= (-1)^n n! \sum_{r=0}^{\left[\frac{n}{2}\right]} \frac{(-1)^r(2x)^{n-2r}}{(n-2r)!r!} \left(\hat{h}^{n-r}e^{-\hat{h}x^2}\right)\eta_0, \quad (2.4.21)$$

thus getting, on account of Eq. (2.4.16),

$$\partial_x^n e^{-\hat{h}x^2}\eta_0$$

$$= (-1)^n n! \sum_{r=0}^{\left[\frac{n}{2}\right]} \frac{(-1)^r(2x)^{n-2r}}{(n-2r)!r!} H_{n-r}(\gamma + 2\,\beta x^2, -\beta)e^{-(\gamma x^2 + \beta x^4)},$$

(2.4.22)

in accordance with

$$\partial_x^n e^{-(\gamma x^2 + \beta x^4)}$$

$$= H_n^{(4)}(-2\,\gamma x - 4\,\beta x^3, -\gamma - 6\,\beta x^2, -4\,\beta x, -\beta)e^{-(\gamma x^2 + \beta x^4)}.$$

(2.4.23)

We have so far described the umbral point of view of the Hermite polynomials. We did not mention yet the higher-order Hermite, which will be touched upon later in this book. We complete our study with the formulation of the theory of Laguerre polynomials in umbral terms.

2.5 Negative Derivative Operator Method and Associated Technicalities

In this section, we present a few useful notions concerning the formalism of negative derivative operators which provide a tool of paramount importance for the umbral/monomiality formulation of Laguerre polynomials. We will reconsider the integration process, by considering a different formulation of the integration by parts [65, 68].

Even classical problems, with well-known solutions, may acquire a different flavor, if viewed within such a perspective which, if properly pursued, may allow further progress disclosing new avenues for their study and generalizations. It is indeed well known that the operation of integration is the inverse of that of derivation; however, such a statement, by itself, does not enable a formalism to establish rules to handle integrals and derivatives on the same footing.

An almost natural environment to place this specific issue is the formalism of real-order derivatives, in which the distinction between integrals and derivatives becomes superfluous. The use of the formalism associated with the fractional-order operators offers new computational tools as, e.g., the extension of the concept of integration by parts. Within such a context we get the following proposition.

Proposition 10. *The integral of a function* $f \in C^\infty$ *can be written in terms of the series* [145]

$$\int_0^x f(\xi)d\xi = \sum_{s=0}^\infty (-1)^s \frac{x^{s+1}}{(s+1)!} f^{(s)}(x), \quad \forall x \in \mathbb{R}, \qquad (2.5.1)$$

where $f^{(s)}(x)$ *denotes the sth-derivative of the integrand function.*

To prove Eq. (2.5.1), we give the following definition.

Definition 9. $\forall x \in \mathbb{R}$, $\forall f \in C^\infty$, let

$$\int_0^x g(\xi)f(\xi)d\xi = {}_0\hat{D}_x^{-1}(g(x)f(x)) \qquad (2.5.2)$$

where

$$_\alpha\hat{D}_x^{-1}s(x) = \int_\alpha^x s(\xi)d\xi, \quad g(x) = 1. \qquad (2.5.3)$$

$_\alpha\hat{D}_x^{-1}$ is the *Negative Derivative Operator (NDO)*.

Proof. We rewrite Eq. (2.5.1) according to Definition 9 and by the use of a slightly generalized form of the *Leibniz formula*, written as

$$_0\hat{D}_x^{-1}(g(x)f(x)) = \sum_{s=0}^{\infty} \binom{-1}{s} g^{(-1-s)}(x)f^{(s)}(x), \qquad (2.5.4)$$

we provide the proof after taking $g(x) = 1$ and after noting that

$$\binom{-1}{s} = (-1)^s, \quad g^{(-1-s)}(x) = \frac{x^{s+1}}{(s+1)!}. \qquad (2.5.5)$$
$\square$

The interesting element of such an analytical tool is that it allows the evaluation of the primitive of a function in terms of an automatic procedure, analogous to that used in the calculus of the derivative of a function. At the same time it marks the conceptual, even though not formal, difference between the two operations. The integrals give rise to a computational procedure involving, most of the times, an infinite number of steps. Equation (2.5.4) becomes useful if, e.g., the function $f(x)$ has peculiar properties under the operation of derivation, like being cyclical, vanishing after a number of steps or other.

The formalism we have just envisaged can be combined, e.g., with the properties of the special polynomials to find useful identities. In the case of two-variable HP, $H_n(x,y)$, satisfying the Properties 4 [18], we obtain the definite integrals in Example 19.

Properties 4. $\forall x, y \in \mathbb{R}$, $\forall n, s \in \mathbb{N} : s \le n$,

$$\partial_x^s H_n(x,y) = \frac{n!}{(n-s)!}H_{n-s}(x,y),$$

$$\partial_y^s H_n(x,y) = \frac{n!}{(n-2s)!}H_{n-2s}(x,y). \qquad (2.5.6)$$

Example 19. $\forall x, y \in \mathbb{R}$, we get

$$\int_0^x H_n(\xi, y)d\xi = \sum_{s=0}^n \frac{(-1)^s \, x^{s+1}}{(s+1)!} \frac{n!}{(n-s)!} H_{n-s}(x, y)$$

$$= \sum_{s=0}^n \frac{x}{(s+1)} \binom{n}{s} (-x)^s H_{n-s}(x, y), \qquad (2.5.7)$$

$$\int_0^y H_n(x, \eta)d\eta = \sum_{s=0}^n \frac{(-1)^s \, y^{s+1}}{(s+1)!} \frac{n!}{(n-2s)!} H_{n-2s}(x, y)$$

and

$$\int_0^x H_n(\xi, y)\cos(\xi)d\xi = \sum_{s=0}^n (-1)^s \frac{\cos\left(x + s\frac{\pi}{2}\right)}{(s+1)!} \frac{n!}{(n-s)!} H_{n-s}(x, y),$$

$$\int_0^y H_n(x, \eta)\cos(\eta)d\eta = \sum_{s=0}^n (-1)^s \frac{\cos\left(y + s\frac{\pi}{2}\right)}{(s+1)!} \frac{n!}{(n-2s)!} H_{n-2s}(x, y).$$

$$(2.5.8)$$

Furthermore, taking into account that (generalization of Eq. (1.2.7)) [18]

$$\partial_x^s \, e^{ax^2 + bx} = H_s(2ax + b, a)e^{ax^2 + bx}, \quad \forall a, b, x \in \mathbb{R} \qquad (2.5.9)$$

and call that $(-1)^n H_n(x, y) = H_n(-x, y)$ [18], we find the further following example.

Example 20.

$$\int_0^x e^{a\xi^2 + b\xi}d\xi = \sum_{s=0}^\infty \frac{x^{s+1}}{(s+1)!} H_s(-2ax - b, a)e^{ax^2 + bx} \qquad (2.5.10)$$

and

$$\int_0^x e^{a\xi^2 + b\xi}\cos(\xi)d\xi = \sum_{s=0}^\infty \frac{\cos\left(x + s\frac{\pi}{2}\right)}{(s+1)!} H_s(-2ax - b, a)e^{ax^2 + bx}.$$

$$(2.5.11)$$

The right-hand side of Eq. (2.5.10) can be viewed as the primitive of the *erfc* function (see Chapter 3).

We can now merge umbral and negative derivative methods to get further results. The use of the properties of the Gaussian functions under repeated derivatives, as, e.g., the generalized form of *HP* $\partial_x^n e^{ax^2} = H_n(2ax, a)e^{ax^2}$ in Eq. (1.2.7), allows the derivation of its *Bessel* umbral counterpart identity (see Chapter 4), as shown in the following.

Example 21. Let $J_0(x)$ be the 0-order Bessel function $\forall x \in \mathbb{R}$, by using Eq. (1.2.17) and the $J_n(x)$-umbral image (4.1.10) (proved in Chapter 4) we get

$$\partial_x^n J_0(x) = \partial_x^n e^{-\hat{c}\left(\frac{x}{2}\right)^2}\varphi_0 = H_n\left(-\hat{c}\frac{x}{2}, -\frac{\hat{c}}{4}\right)e^{-\hat{c}\left(\frac{x}{2}\right)^2}\varphi_0$$

$$= (-1)^n n! \sum_{r=0}^{\lfloor\frac{n}{2}\rfloor} \frac{(-1)^r x^{n-2r}\hat{c}^{n-r}}{2^{n-2r}2^r(n-2r)!r!} \sum_{s=0}^{\infty}\frac{(-1)^s\left(\frac{x}{2}\right)^{2s}\hat{c}^s}{s!}\varphi_0$$

$$= (-1)^n n! \sum_{r=0}^{\lfloor\frac{n}{2}\rfloor} \frac{(-1)^r\left(\frac{x}{2}\right)^{-r}}{2^r(n-2r)!r!} \sum_{s=0}^{\infty}\frac{(-1)^s\left(\frac{x}{2}\right)^{2s+(n-r)}}{s!(s+(n-r))!}$$

$$= (-1)^n n! \sum_{r=0}^{\lfloor\frac{n}{2}\rfloor} \frac{(-2)^{-r}x^{-r}}{r!(n-2r)!}J_{n-r}(x). \qquad (2.5.12)$$

We can therefore *translate* the identity (2.5.10) as

$$\int_0^x J_0(\xi)d\xi = \sum_{s=0}^{\infty}\frac{x^{s+1}}{(s+1)!}\left(s!\sum_{r=0}^{\lfloor\frac{n}{2}\rfloor}\frac{(-2)^{-r}x^{-r}}{r!(s-2r)!}\right)J_{s-r}(x). \qquad (2.5.13)$$

Further integral transforms can be framed within the same context.

Example 22. The well-known identities [2] $\forall x \in \mathbb{R}^+$

$$\int_0^{\infty} J_0\left(2\sqrt{xu}\right)\sin(u)du = \cos(x),$$

$$\int_0^{\infty} J_0\left(2\sqrt{xu}\right)\cos(u)du = \sin(x),$$

$$(2.5.14)$$

can be proved by following the method illustrated so far. They can indeed be easily stated as follows:

$$\int_0^\infty J_0\left(2\sqrt{xu}\right)e^{iu}du = \int_0^\infty e^{-(\hat{c}x-i)u}du\,\varphi_0 = \frac{1}{\hat{c}x-i}\varphi_0$$

$$= i\sum_{r=0}^\infty(-i\hat{c}x)^r\varphi_0 = -ie^{-ix}. \qquad (2.5.15)$$

Example 23. Reminding Lemma 2, which provides $J_0\left(2\sqrt{x}\right) = C_0(x) = \sum_{r=0}^\infty\frac{(-x)^r}{r!^2}$, where $C_0(x)$ is the 0-order Tricomi–Bessel function, satisfying the identity $\forall x\in\mathbb{R}, \forall s\in\mathbb{R}_0^+$ [166]

$$\partial_x^s C_0(x) = (-1)^s C_s(x), \quad C_s(x) = \sum_{r=0}^\infty\frac{(-x)^r}{r!(r+s)!}, \qquad (2.5.16)$$

we find

$$\int_0^\infty u^s C_s(xu)\sin(u)du = (-1)^s\cos\left(x+s\frac{\pi}{2}\right),$$
$$\int_0^\infty u^s C_s(xu)\cos(u)du = (-1)^s\sin\left(x+s\frac{\pi}{2}\right) \qquad (2.5.17)$$

and, by a straightforward application of Eqs. (1.2.17)–(1.2.20), we also obtain

$$\int_0^\infty C_0(xu)J_0(u)du = J_0(x). \qquad (2.5.18)$$

In the forthcoming section, we use the same formalism to introduce the *Laguerre polynomial* (LP) families.

2.6 Laguerre Polynomials and Relevant Umbral Forms

In this section, we see how the different concepts and computational tools, developed in the previous sections, can be merged to provide a "non-standard" point of view to the theory of the two-variable *LPs*, defined by the series (see [47] and references therein)

$$L_n(x,y) = n! \sum_{r=0}^{n} \frac{(-1)^r y^{n-r} x^r}{(n-r)! r!^2}, \quad \forall x, y \in \mathbb{R}, \ \forall n \in \mathbb{N}. \qquad (2.6.1)$$

Proposition 11. *On account of the fact that, by Definition 9,*

$$_0\hat{D}_x^{-r} 1 = \frac{x^r}{r!}, \quad \forall x \in \mathbb{R}_0^+, \qquad (2.6.2)$$

we can cast Eq. (2.6.1) in the form

$$L_n(x,y) = n! \sum_{r=0}^{n} \frac{(-1)^r y^{n-r} {}_0\hat{D}_x^{-r}}{(n-r)! r!} = \left(y - {}_0\hat{D}_x^{-1} \right)^n \qquad (2.6.3)$$

(the 1 on the right side of the operator has been omitted for brevity). The LP have therefore been transformed into a Newton binomial involving a negative derivative operator.

Properties 5. It is fairly natural to infer from the operational definition in Eq. (2.6.3) that

(a)

$$\left(y - {}_0\hat{D}_x^{-1} \right) \left(y - {}_0\hat{D}_x^{-1} \right)^n = \left(y - {}_0\hat{D}_x^{-1} \right)^{n+1}, \qquad (2.6.4)$$

(b) Since $_0\hat{D}_x^{-1}\partial_x = \hat{1}$, it[g] is easily checked by recursivity that

$$-\partial_x x \partial_x \left(y - {}_0\hat{D}_x^{-1} \right)^n = n L_{n-1}(x,y) \qquad (2.6.5)$$

and that

$$\left[-\partial_x x \partial_x, y - {}_0\hat{D}_x^{-1} \right] = 1. \qquad (2.6.6)$$

Proposition 12. *According to Definition 6, Eqs. (2.6.4)–(2.6.5) define the multiplicative and derivative operators of LP, namely*

$$\hat{M} = y - \hat{D}_x^{-1}, \qquad (2.6.7)$$

$$\hat{P} = -\partial_x x \partial_x. \qquad (2.6.8)$$

---

[g]It must be stressed that the condition holds if the operator $_0\hat{D}_x^{-1}\partial_x$ acts on polynomials (note that the lower integration limit is zero) or on a function vanishing at zero.

In particular, Eq. (2.6.8) is referred, in the relevant mathematical literature, as Laguerre derivative $_L\partial_x$.

A number of consequences can be drawn from the previous formalism.

Properties 6. $\forall x, y \in \mathbb{R}, \forall n \in \mathbb{N}$,

(i) **Differential equation satisfied by Laguerre operators**

$$- \left(\left(y - \hat{D}_x^{-1} \right) \partial_x x \partial_x \right) L_n(x, y) = n L_n(x, y), \qquad (2.6.9)$$

yields

$$y \, x \, Z'' + (y - x) Z' + n Z = 0, \quad Z = L_n(x, y). \qquad (2.6.10)$$

Even though not explicitly mentioned, the two-variable Laguerre reduces to their ordinary counterpart [3] for $y = 1$.

(ii) **Laguerre Heat Equation (LHE)**
The following identity naturally follows from the previous identities [18]:

$$\begin{cases} \partial_y L_n(x, y) = -\partial_x x \partial_x L_n(x, y) \\ L_n(x, 0) = \dfrac{(-x)^n}{n!}. \end{cases} \qquad (2.6.11)$$

It is evident that Eq. (2.6.11) provides the further operational definition of LP

$$L_n(x, y) = e^{-y \partial_x x \partial_x} \frac{(-x)^n}{n!}, \qquad (2.6.12)$$

which can be checked after the use of the (surprising) identity

$$(\partial_x x \partial_x)^s = \partial_x^s x^s \partial_x^s, \qquad (2.6.13)$$

in turn proved by induction.

Equation (2.6.11) is an initial value problem analogous to the heat equation satisfied by the HP, this justifies its definition of LHE. In Example 24, we will discuss relevant solutions involving different initial conditions.

(iii) **Associated generating function(s) as straightforward consequence of the envisaged formalism**

We find indeed

$$\sum_{n=0}^{\infty} t^n L_n(x,y) = \frac{1}{1-\left(y-\hat{D}_x^{-1}\right)t} = \frac{1}{1-yt}\left(\frac{1}{1+\frac{t\hat{D}_x^{-1}}{1-yt}}\right)$$

$$= \frac{1}{1-yt}\sum_{r=0}^{\infty}\frac{(-1)^r}{(1-yt)^r}\hat{D}_x^{-r}. \qquad (2.6.14)$$

Finally, by the use of Eq. (2.6.2), we find

$$\sum_{n=0}^{\infty} t^n L_n(x,y) = \frac{1}{1-yt}e^{-\frac{xt}{1-yt}}, \qquad |\,yt\,|<1. \qquad (2.6.15)$$

Other generating functions are obtained by combining elementary algebraic rules and those associated with the negative derivative operator, like, e.g.,

$$\sum_{n=0}^{\infty} t^n L_n(x,y) = \sum_{n=0}^{\infty} t^n\left(y-\hat{D}_x^{-1}\right)^n = e^{yt}e^{-t\hat{D}_x^{-1}}$$

$$= e^{yt}\sum_{r=0}^{\infty}\frac{(-t)^r}{r!}\hat{D}_x^{-r}, \qquad (2.6.16)$$

thus allowing the conclusion

$$\sum_{n=0}^{\infty} t^n L_n(x,y) = e^{yt}\sum_{r=0}^{\infty}\frac{(-tx)^r}{r!^2} = e^{yt}C_0(xt). \qquad (2.6.17)$$

Example 24. Let us now apply the previous results to the solution of the *LHE*

$$\begin{cases} \partial_y F(x,y) = -\partial_x x \partial_x F(x,y) \\ F(x,0) = e^{-x}. \end{cases} \qquad (2.6.18)$$

The use of the evolution operator method yields

$$F(x,y) = \frac{1}{1-y}e^{-\frac{x}{1-y}}, \qquad (2.6.19)$$

which has a limited convergence radius, fixed by $|\,y\,|<1$.

In the case of a generic initial function $F(x,0) = g(x)$, the solution of the *LHE* reads (see Eq. (7.4.17) for the relevant proof)

$$F(x,y) = e^{\frac{x}{y}} \int_0^\infty e^{-s} C_0 \left(\frac{x}{y}s\right) g(-xs)ds, \qquad (2.6.20)$$

which plays the same role of the Gauss–Weierstrass transform and can be exploited to infer the *LP* orthoganal properties.

Before proceeding further we comment on the geometrical properties of *LP*, by taking advantage of Eqs. (2.6.11), the procedure is analogous to that already discussed for *HP* and we find indeed what is shown in Fig. 2.2.

The previous formulas suggest geometrical representations for the two-variable *LP*, which are displayed in the graphics. The exponential operator transforms an ordinary monomial $\frac{(-x)^n}{n!}$ into a Laguerre type polynomial. The monomial-to-polynomial transition is shown by moving the cutting plane orthogonal to the y axis. For a specific value of the polynomial degree n, the polynomials lie on the cutting $L_n(x,y)$ plane.

It is evident that both for Laguerre and Hermite polynomials the exponential operator acting on the monomial is a kind of shift in the y variable and indeed we have

$$e^{-y_2 \partial_x x \partial_x} L_n(x, y_1) = L_n(x, y_1 + y_2),$$
$$e^{y_2 \partial_x^2} H_n(x, y_1) = H_n(x, y_1 + y_2). \qquad (2.6.21)$$

Furthermore, to recover monomials from *HP* and *LP* it will be enough to set

$$e^{y \partial_x x \partial_x} L_n(x, y) = L_n(x, 0) = \frac{(-x)^n}{n!},$$
$$e^{-y \partial_x^2} H_n(x, y) = H_n(x, 0) = x^n. \qquad (2.6.22)$$

In the forthcoming section, we discuss the transition toward the Umbral treatment.

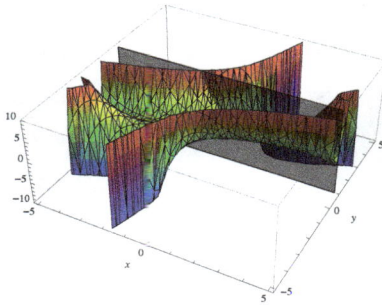

(a) $L_4(x, y)$ be cut by the plane $y = 1$.

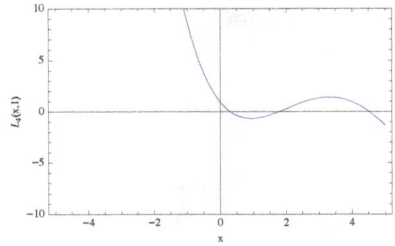

(b) $L_4(x, 1)$.

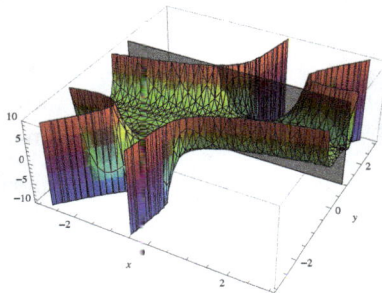

(c) $L_8(x, y)$ be cut by the plane $y = 1$.

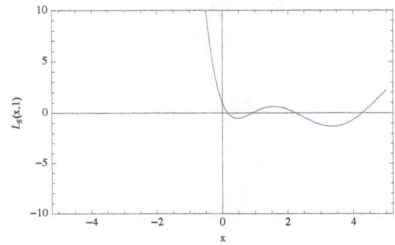

(d) $L_8(x, 1)$.

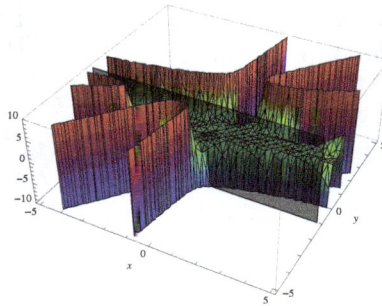

(e) $L_9(x, y)$ be cut by the plane $y = -1$.

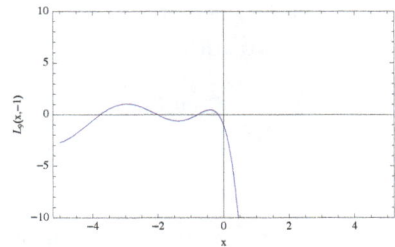

(f) $L_9(x, -1)$.

Figure 2.2:   Geometrical representation of two-variable Laguerre polynomials in 3D and 2D, for different n and y values [9].

2.7 Umbral Version of Laguerre and Hermite Associated Polynomials

Most of the results we have derived in the previous section can be deduced using the umbral formalism as its results are evident in the following.[h]

Proposition 13. $\forall x, y \in \mathbb{R}$, $\forall n \in \mathbb{N}$, *by using Eq.* (1.1.6) *and the same procedure of Proposition* 8, *we provide*

$$L_n(x, y) = (y - \hat{c}x)^n \varphi_0. \tag{2.7.1}$$

The use of the previous proposition allows a significant simplification of the theory of LP and Eq. (2.7.1) will be largely exploited in the forthcoming chapters. We note that such a point of view has opened new avenues in the derivation of *Lacunary Generating Functions* [16] and for the relevant combinatorial interpretation [163].

Let us note that Eq. (2.7.1) suggests that the definition can be extended to negative real values of the index. In this case, we will not deal with polynomials any more but with Laguerre functions.

Proposition 14. *We provide the integral form of negative-order LP*

$$L_{-\nu}(x, y) = \frac{1}{\Gamma(\nu)} \int_0^\infty s^{\nu-1} e^{-sy} C_0(-sx) ds, \quad \forall \nu \in \mathbb{R}^+, \ \forall x, y \in \mathbb{R}. \tag{2.7.2}$$

Proof. By applying Laplace transform and Eq. (1.2.20), we get

$$L_{-\nu}(x, y) = (y - \hat{c}x)^{-\nu} \varphi_0 = \frac{1}{\Gamma(\nu)} \int_0^\infty s^{\nu-1} e^{-sy} e^{\hat{c}xs} \varphi_0 ds$$

$$= \frac{1}{\Gamma(\nu)} \int_0^\infty s^{\nu-1} e^{-sy} C_0(-sx) ds, \tag{2.7.3}$$

$\square$

---

[h]We note that in a similar way we can write from Eq. (1.1.6)

$$L_n(x, y) = \sum_{s=0}^n \binom{n}{s} y^{n-s} (-\hat{c}x)^s \qquad \varphi_0 = \sum_{s=0}^n \frac{(-1)^s n!}{(s!)^2 (n-s)!} y^{n-s} x^s.$$

Corollary 9. *In a similar way, by applying identity (1.2.18), we can recast (2.7.2) as*

$$L_{-\nu}(x,y) = \frac{1}{\Gamma(\nu)} \int_0^\infty s^{\nu-1} e^{-sy} J_0\left(2\sqrt{-xs}\right) ds. \qquad (2.7.4)$$

It has been shown in [3] that the negative-order LP satisfy the same recurrences and differential equation of their polynomial counterpart.

Example 25. It is evident that the derivation of integrals of the type

$$\int_{-\infty}^\infty e^{-ax^2} J_0(bx)dx = \int_{-\infty}^\infty e^{-x^2\left(a+\frac{b^2}{4}\hat{c}\right)} dx\, \varphi_0 = \sqrt{\pi}\left(a+\frac{b^2}{4}\hat{c}\right)^{-\frac{1}{2}} \varphi_0$$

$$= \sqrt{\pi} L_{-\frac{1}{2}}\left(-\frac{b^2}{4}, a\right) \qquad (2.7.5)$$

are a straightforward consequence of the previous formalism.[i]

A further point deserving attention is the definition of *Associated Laguerre polynomials*, which within the present framework are introduced as

Definition 10. We give a slightly different definition of the two-variable Associated Laguerre polynomials, $\forall x, y \in \mathbb{R}$, $\forall n \in \mathbb{N}$, $\forall \alpha \in \mathbb{R}$,

$$\Lambda_n^{(\alpha)}(x,y) := \hat{c}^\alpha(y-\hat{c}x)^n \varphi_0 = n! \sum_{r=0}^n \frac{(-1)^r y^{n-r} x^r}{(n-r)!r!\Gamma(r+\alpha+1)}, \qquad (2.7.6)$$

linked to the ordinary definition[j] [3] by

$$L_n^{(\alpha)}(x,y) = \frac{\Gamma(n+\alpha+1)}{n!} \Lambda_n^{(\alpha)}(x,y). \qquad (2.7.7)$$

---

[i]Equation (2.7.5) will be interpreted in terms of two-variable generalized Bessel functions in Chapters 3 and 4.

[j]Ordinary Associated Hermite polynomials can be represented by

$$L_n^{(m)}(x,y) = (n+m)! \sum_{r=0}^\infty \frac{(-1)^r y^{n-r} x^r}{(n-r)!r!(r+m)!}.$$

Corollary 10. *The structural similarity between HP and LP due to the "unifying" formalism developed so far, suggests (see Eqs. (2.3.2)–(2.3.3)) the definition of Associated Hermite polynomials as*

$$H_n(x, y \mid p) = {}_y\hat{h}^p \left(x + {}_y\hat{h} \right)^n \theta_0$$

$$= n! \sum_{r=0}^{n} \frac{(r+p)! \, y^{\frac{r+p}{2}} \, x^{n-r}}{\Gamma \left(\frac{r+p}{2} + 1 \right) (n-r)! r!} \left| \cos \left((r+p)\frac{\pi}{2} \right) \right|.$$

$$(2.7.8)$$

According to the previous definition, we can state the following corollary.

Corollary 11. *We provide the index-duplication formula*

$$H_{2n}(x, y) = \left(x + {}_y\hat{h} \right)^n \left(x + {}_y\hat{h} \right)^n \theta_0$$

$$= n! \sum_{s=0}^{n} \frac{x^{n-s}}{(n-s)! s!} y \hat{h}^s \left(x + {}_y\hat{h} \right)^n = n! \sum_{s=0}^{n} \frac{x^{n-s}}{(n-s)! s!} H_n(x, y \mid s)$$

$$(2.7.9)$$

and argument duplication formula

$$H_n(2x, y) = \left[\left(x + \frac{y\hat{h}}{2} \right) + \left(x + \frac{y\hat{h}}{2} \right) \right]^n \theta_0$$

$$= \sum_{s=0}^{n} \binom{n}{s} \sum_{r=0}^{s} \binom{s}{r} \frac{x^r}{2^{s-r}} H_{n-s} \left(x, \frac{y}{4} \mid s - r \right). \quad (2.7.10)$$

It is furthermore easily checked that

$$x^n = \left[\left(x + {}_y\hat{h} \right) - {}_y\hat{h} \right]^n \theta_0 = \sum_{r=0}^{n} \binom{n}{r} (-1)^r H_{n-r}(x, y \mid r) \quad (2.7.11)$$

or similarly

$$x^n = \left[\left(x + {}_y\hat{h} \right) - {}_y\hat{h} \right]^n \theta_0 = \sum_{r=0}^{n} \binom{n}{r} (-1)^{n-r} H_r(x, y \mid n - r)$$

$$(2.7.12)$$

and that

$$H_{n+m}(x,y) = \left(x + {}_y\hat{h}\right)^m \left(x + {}_y\hat{h}\right)^n \theta_0 = \sum_{r=0}^{m} \binom{m}{r} x^{m-r} H_n(x,y \mid r).$$

(2.7.13)

The last identity is a reformulation of the *Nilsen* theorem, concerning the sum of the indices of *HP* [86]. The previous results (2.7.10)–(2.7.13) occur in the literature in different forms [86].

Even though not explicitly mentioned, the Hermite umbra can be raised to *any real power* and this allows noticeable freedom in guessing possible generalizations. A fairly direct example is provided by the following extension.

Lemma 9. $\forall \alpha, \beta \in \mathbb{R}$

$$H_n(x,y \mid \beta; \alpha) = {}_y\hat{h}^\beta \left(x + {}_y\hat{h}^\alpha\right)^n \theta_0,$$

(2.7.14)

yielding a family of polynomials with generating function

$$\sum_{n=0}^{\infty} \frac{t^n}{n!} H_n(x,y \mid \beta; \alpha) = e^{xt} y^{\frac{\beta}{2}} e_{\alpha,\beta}\left(y^{\frac{\alpha}{2}} t\right),$$

$$e_{\alpha,\beta}(x) = \sum_{r=0}^{\infty} \frac{\Gamma(\alpha r + \beta + 1) x^r}{\Gamma\left(\frac{\alpha r + \beta}{2} + 1\right) r!} \left| \cos\left(\frac{\alpha r + \beta}{2}\pi\right) \right|.$$

(2.7.15)

Their properties can easily be studied and they are framed within the context of the Sheffer family. They can accordingly be defined through the operational rule

$$H_n(x,y \mid \beta; \alpha) = y^{\frac{\beta}{2}} e_{\alpha,\beta}\left(y^{\frac{\alpha}{2}} \partial_x\right) x^n.$$

(2.7.16)

We have collected enough material and developed sufficient confidence with the umbral formalism technicalities, to apply it to other families of polynomials.

2.8 Appéll and Sheffer Polynomials: Umbral and Monomiality Points of View

We have described Hermite and Laguerre polynomials' properties, by using a procedure that has revealed particular elements of simplicity, displaying features that can be extended to other families of polynomials.

Our description has proceeded through two entangled steps:

(1) the definition of a set of operators, realizing a Weyl algebra and playing the role of derivative and multiplicative operators;
(2) the introduction of an Umbral picture.

The characterizing feature of both steps of the description is that these families of polynomials can be represented as ordinary monomials. The extension of this point of view to other families of polynomials is the central topic of this chapters introductory section.

The Appéll set [6] is a natural benchmark to accomplish such a task. It allows a natural umbral (monomial) treatment, that will be introduced along the prescription discussed for Hermite (and Laguerre as well) polynomial families. Before treating the problem in its generality, we start our discussion by presenting a simple example.

Example 26. The polynomials of the truncated exponential (which will be treated in Eq. (6.2.12) and the following in a slightly different form[k]) are defined through the generating function:

$$\sum_{n=0}^{\infty} \frac{t^n}{n!} \tilde{e}_n(x) = \frac{e^{xt}}{1-t}, \qquad |\,t\,| < 1, \qquad (2.8.1)$$

which can be written in umbral form as

$$\sum_{n=0}^{\infty} \frac{t^n}{n!} \tilde{e}_n(x) = e^{(x+\hat{e})t} \iota_0, \qquad \hat{e}^r \iota_0 = r!. \qquad (2.8.2)$$

---

[k]In Eq. (6.2.12), we will use the representation $e_n(x) = \sum_{r=0}^{n} \frac{x^r}{r!}$, in the present context $\tilde{e}_n(x) = n! e_n(x)$.

It yields the relevant expression in the form of a series, namely

$$\tilde{e}_n(x) = (x + \hat{e})^n \iota_0 = \sum_{s=0}^{n} \binom{n}{s} x^{n-s} \hat{e}^s \iota_0 = \sum_{s=0}^{n} \frac{n!}{(n-s)!} x^{n-s}$$

$$= e^x \Gamma(n+1, x) \tag{2.8.3}$$

which verifies the sum of the series $\tilde{e}_n(x) = n! \sum_{r=0}^{n} \frac{x^r}{r!}$.

The monomial nature of the polynomials is also easily stated. We find indeed from Eqs. (2.8.2)–(2.8.3) the following identities:

$$\partial_x \tilde{e}_n(x) = n\, \tilde{e}_{n-1}(x), \quad \tilde{e}_{n+1}(x) = \left(x + \frac{1}{1 - \partial_x} \right) \tilde{e}_n(x). \tag{2.8.4}$$

The first is an obvious consequence of the umbral definition in Eq. (2.8.3)

$$\partial_x \tilde{e}_n(x) = n(x + \hat{e})^{n-1} \iota_0. \tag{2.8.5}$$

The second requires a few comments more. We note indeed that

$$\tilde{e}_{n+1}(x) = (x + \hat{e})(x + \hat{e})^n \iota_0 \tag{2.8.6}$$

which can be exploited to find the generating function

$$\sum_{n=0}^{\infty} \frac{t^n}{n!} \tilde{e}_{n+1}(x) = (x + \hat{e}) e^{(x+\hat{e})t} \iota_0 = \left(x\, e^{(x+\hat{e})t} + e^{xt} \partial_t e^{\hat{e}t} \right) \iota_0$$

$$= e^{xt} (x + \partial_t)\, e^{\hat{e}t} \iota_0 = e^{xt} (x + \partial_t) \sum_{r=0}^{\infty} t^r$$

$$= \left(x + \frac{1}{1-t} \right) \frac{e^{xt}}{1-t}. \tag{2.8.7}$$

Written in a more convenient (for our purposes) form, recalling the operational method (2.2.1) (see Chapter 7, Exercise 52), we obtain

$$\sum_{n=0}^{\infty} \frac{t^n}{n!} \tilde{e}_{n+1}(x) = \left(x + \frac{1}{1 - \partial_x} \right) \sum_{n=0}^{\infty} \frac{t^n}{n!} \tilde{e}_n(x) \tag{2.8.8}$$

which, eventually, yields the second of Eq. (2.8.4), after equating the same like t-powers.

Example 27. A slight extension of the procedure leading to the Eqs. (2.8.7)–(2.8.8) can be exploited to derive the generating function

$$\sum_{n=0}^{\infty} \frac{t^n}{n!} \tilde{e}_{n+l}(x) = (x + \hat{e})^l e^{(x+\hat{e})t} {}_\iota 0 = \sum_{r=0}^{l} \binom{l}{r} x^{l-r} \hat{e}^r e^{(x+\hat{e})t} {}_\iota 0$$

$$= e^{xt} \sum_{r=0}^{l} \binom{l}{r} x^{l-r} \partial_t^r e^{\hat{e}t} {}_\iota 0 = e^{xt} \sum_{r=0}^{l} \binom{l}{r} x^{l-r} \frac{r!}{(1-t)^{r+1}}$$

$$= \frac{e^{xt} x^l \, l!}{1-t} \sum_{r=0}^{l} \frac{1}{(l-r)!} x^{-r} \frac{1}{(1-t)^r}$$

$$= \frac{e^{xt} x^l \, l!}{1-t} \frac{e^{x-tx} \Gamma(l+1, x-tx)}{(x-tx)^l l!} = \frac{e^x \Gamma(l+1, x(1-t))}{(1-t)^{l+1}}$$

$$= \frac{e^{xt}}{(1-t)^{l+1}} \tilde{e}_l(x(1-t)). \tag{2.8.9}$$

Proposition 15. *According to the previous remarks, the polynomials of the truncated exponential are monomials under the action of the operators*

$$\hat{M} = x + \frac{1}{1 - \partial_x}, \quad \hat{P} = \partial_x. \tag{2.8.10}$$

The differential equation satisfied by $\tilde{e}_n(x)$ can therefore be derived from the algebraic condition $\hat{M}\hat{P} = n$, which yields, by applying Eqs. (2.8.4),

$$\left(x + \frac{1}{1 - \partial_x}\right) \partial_x \tilde{e}_n(x) = n \, \tilde{e}_n(x). \tag{2.8.11}$$

Corollary 12. *Equation (2.8.11) can be written in different ways (see Chapter 7) and here we note the following integro-differential form (whose meaning will be commented on later in this chapter):*

$$x \, \partial_x \, e_n(x) + \partial_x \int_0^{\infty} e^{-s} e_n(x+s) ds = n \, e_n(x), \tag{2.8.12}$$

obtained by the use of standard Laplace transform techniques.

In more general terms, the Appéll polynomials are characterized by the generating function [157]

$$\sum_{n=0}^{\infty} \frac{t^n}{n!} a_n(x) = A(t)e^{xt}. \qquad (2.8.13)$$

The umbral and monomiality properties are easily established by a straightforward extension of the previous procedure, which yields the following derivative and multiplicative operators (see Chapter 7 for details):

$$\hat{M} = x + \frac{A'(\partial_x)}{A(\partial_x)}, \qquad A'(\xi) = \partial_\xi A(\xi), \qquad \hat{P} = \partial_x. \qquad (2.8.14)$$

Hermite polynomials belong to the Appéll family, with $A(t) = e^{yt^2}$. It is a simple matter to derive the relevant multiplicative operator from the first of Eq. (2.8.14).

The next point we discuss is about the Sheffer sequences [153], a generalization of the Appéll counterpart, characterized by the generating function

$$\sum_{n=0}^{\infty} \frac{t^n}{n!} s_n(x) = g\left(f^{-1}(t)\right) e^{f^{-1}(t)x}. \qquad (2.8.15)$$

The proof of the relevant monomiality is easily achieved and shown as follows.

Example 28. We first note that

$$\sum_{n=0}^{\infty} \frac{t^n}{n!} f(\partial_x) s_n(x) = f(\partial_x) g\left(f^{-1}(t)\right) e^{f^{-1}(t)x}$$

$$= f\left(f^{-1}\right)(t) g\left(f^{-1}(t)\right) e^{f^{-1}(t)x} = tg\left(f^{-1}(t)\right) e^{f^{-1}(t)x} \qquad (2.8.16)$$

or

$$\sum_{n=0}^{\infty} \frac{t^n}{n!} f(\partial_x) s_n(x) = \sum_{m=0}^{\infty} \frac{t^{m+1}}{m!} f(\partial_x) s_m(x) \qquad (2.8.17)$$

which, after confronting the two series on both sides of Eq. (2.8.17), yields the identity

$$f(\partial_x)s_n(x) = ns_{n-1}(x). \tag{2.8.18}$$

According to Eq. (2.8.18), we can state that Sheffer's sequence monomial derivative is

$$\hat{P} = f(\partial_x). \tag{2.8.19}$$

Regarding its multiplicative counterpart, we proceed as for the Appéll case. We first assume that the following umbral definition is allowed:

$$s_n(x) = (x + \hat{s})^n \varsigma_0, \quad \hat{s}^n \varsigma_0 = s_n. \tag{2.8.20}$$

We can therefore write

$$\sum_{n=0}^{\infty} \frac{t^n}{n!} s_{n+1}(x) = (x + \hat{s})e^{(x+\hat{s})t} \varsigma_0 \tag{2.8.21}$$

and note that it can be rearranged as

$$\sum_{n=0}^{\infty} \frac{t^n}{n!} s_{n+1}(x) = \partial_t e^{(x+\hat{s})t} \varsigma_0 = \partial_t [A\left(f^{-1}(t)\right) e^{xf^{-1}(t)}]. \tag{2.8.22}$$

After setting

$$s_{n+1(x)=\hat{M}} s_n(x) \tag{2.8.23}$$

we can conclude (for the computational details see Chapter 7)

$$\hat{M} = \left(x + \frac{A'(\partial_x)}{A(\partial_x)}\right) \frac{1}{f'(\partial_x)}, \quad f'(\xi) = \partial_\xi f(\xi). \tag{2.8.24}$$

The differential equation satisfied by the Sheffer polynomials is accordingly given by

$$\left(g(\partial_x)x + g'(\partial_x)\right) \frac{f(\partial_x)}{f'(\partial_x)} s_n(x) = ng(\partial_x)s_n(x). \tag{2.8.25}$$

The question whether the monomial treatment can be extended to any polynomial family will be touched on later in this chapter. Specific examples of application can be found in the forthcoming section and in Chapter 7.

2.9   Genesis of the Monomiality Principle

Before passing to non-Appéll/Sheffer" polynomial families, and explaining how they can be viewed within the umbral formalism, we like to add a few comments regarding the matter treated so far and the genesis of the Monomiality Principle. There is a kind of mantra, opening many of the papers concerning the theory and the application of monomiality:

> *This definition (of monomials) traces back to a paper by J.F. Steffensen who introduced the concept of "poweroid" ..., recently improved by .... and widely used in several applications...*

In absence of an appropriate exposition of the point of view developed by Steffensen [160–162] the above statement is not particularly significant. The comments, which follow, are aimed at clarifying how the two points of view should be confronted. We introduce the Steffensen important intuition, using a recent interesting paper by Dowker [89].

The mathematical environment, in which the concept of poweroid harbored, was that of finite differences, a branch of calculus treated in symbolic terms since its foundations at the beginning of 17th century [89]. Within this context, along with the difference operator, the most commonly exploited tool was that of factorial. The latter played, for the calculus of differences, the same role of the power. We start by a simple example to clarify what we mean.

Example 29. We introduce the operator

$$\hat{P} = e^{\partial_x} - 1. \tag{2.9.1}$$

Its action on a continuous infinitely differentiable function $f(x)$ yields

$$\hat{P}^n f(x) = \sum_{s=0}^{n} (-1)^s \binom{n}{s} e^{(n-s)\partial_x} f(x) = \sum_{s=0}^{n} (-1)^s \binom{n}{s} f(x+n-s).$$

$$\tag{2.9.2}$$

Along with the operator in Eq. (2.9.1), we *construct* its counterpart, realizing with $\hat{P}$ a Weyl algebra. It is easily checked that [14]

$$\hat{M} = xe^{-\partial_x} \tag{2.9.3}$$

satisfies the condition $[\hat{P}, \hat{M}] = 1$. If we assume that (2.9.1)–(2.9.2) are the derivative and multiplicative operators of a not yet specified monomial set $p_n(x)$, we can define such a set by the use of the condition (implicitly assuming $p_0(x) = 1$)

$$p_n(x) = \left(xe^{-\partial_x}\right)^n 1. \tag{2.9.4}$$

The explicit form of the associated monomials are then obtained by iteration, namely

$$p_n(x) = \left(xe^{-\partial_x}\right) \cdots \left(xe^{-\partial_x}\right) 1 = (x-n)(x-(n-1)) \cdots x. \tag{2.9.5}$$

Upon introducing the Pochhammer symbol $(x)_n$, we set

$$p_n(x) = (x)_n = \frac{\Gamma(x+1)}{\Gamma(x-n+1)} = \phi_n(x) \tag{2.9.6}$$

and we can check that

$$\hat{P}(x)_n = n(x)_{n-1}, \quad \hat{M}(x)_n = (x)_{n+1}. \tag{2.9.7}$$

It is easily seen that the polynomials $\phi_n(x)$, now recognized as *factorial polynomials*, satisfy the eigenvalue problem [14]

$$x\left(\phi_n(x) - \phi_n(x-1)\right) = n\phi_n(x). \tag{2.9.8}$$

The quantity is what Steffensen called *poweroid* [160–162]. It should be stressed that such a definition was limited to the first of Eq. (2.9.7) (namely to factorial polynomials) and no notion of multiplicative operator was introduced in his study.

The definition of "derivative" operator and of poweroid as well, given by Steffensen, is more general and those in Eqs. (2.9.1)–(2.9.6) are a particular case in point. Using a different notation from that exploited in [89]

$$\hat{P}_{\alpha,\beta} = \frac{e^{\alpha \partial_x} \left(e^{\beta \partial_x} - 1 \right)}{\beta}. \tag{2.9.9}$$

The corresponding poweroid is constructed from the operational identity (implicitly assuming)

$$p_n(x; \alpha, \beta) = x e^{-n\alpha \partial_x} \left(x e^{-\beta \partial_x} \right)^n$$

$$= x(x - n\alpha - \beta)(x - n\alpha - 2\beta) \cdots (x - n\alpha - (n-1)\beta). \tag{2.9.10}$$

It is a straightforward exercise to prove that $\hat{P}_{\alpha,\beta}$ is a derivative operator for the poweroid (2.9.10)

$$\hat{P}_{\alpha,\beta} p_n(x; \alpha\beta) = n p_{n-1}(x; \alpha, \beta). \tag{2.9.11}$$

No multiplicative operator has been introduced for this case, too. This point will be further commented on a dedicated section in Chapter 7.

The concept of poweroid was contained in earlier works, tracing back to Jeffrey [107, 108], Boole [32] and others (for a more general account, see [89]) and to the more recent paper on umbral calculus, where they have been indicated as "basic polynomials" (Sheffer and Rota) [153] and "associated polynomials" (Roman and Rota) [89, 147, 149–151].

Further comments on these points will be clarified in a dedicated section in Chapter 7, where we will present a more refined formulation of the monomiality principle.

Chapter 3

Special Polynomials and Umbral Operators

In this chapter, we underline the deep link which connects the umbral operators with a wide family of orthogonal polynomials. We provide a wide range of integral or PDE results in terms of special polynomials easily treatable if in umbral forms.

3.1 An Umbral Treatment of Gegenbauer, Legendre and Jacobi Polynomials

Special polynomials, ascribed to the family of Gegenbauer, Legendre and Jacobi and of their associated forms, can be expressed in an operational way, which allows a high degree of flexibility for the formulation of the relevant theory [60].

In the following, we develop the study of the properties of the Gegenbauer polynomials [3] and of their generalized forms in terms of umbral operators. We introduce the main concepts, associated with the technique we are going to deal with, starting from a very simple example.

Example 30. We consider $\forall x \in \mathbb{R} : x > -1$ the elementary function

$$e^{(\nu)}(x) = (1+x)^{-\nu}, \qquad \mathrm{Re}(\nu) > 0. \qquad (3.1.1)$$

According to the use of standard Laplace transform identities (1.3.4), the function (3.1.1) can be rewritten as

$$e^{(\nu)}(x) = \frac{1}{\Gamma(\nu)} \int_0^\infty e^{-s} s^{\nu-1} e^{-sx} ds. \qquad (3.1.2)$$

By following [54], we use the operational rule $(\alpha)^{x\,\partial_x} f(x) = f(\alpha\,x)$ in (1.1.16) to write

$$e^{(\nu)}(x) = \frac{1}{\Gamma(\nu)} \int_0^\infty e^{-s} s^{\nu-1} s^{x\,\partial_x} e^{-x} ds. \qquad (3.1.3)$$

We can therefore recast the function (3.1.1) as

$$\frac{1}{\Gamma(\nu)} \int_0^\infty e^{-s} s^{\nu-1} s^{x\,\partial_x} e^{-x} ds = \left(\frac{1}{\Gamma(\nu)} \int_0^\infty e^{-s} s^{\nu-1} s^{x\,\partial_x} ds \right) e^{-x}$$

$$= \hat{\Gamma}_\nu\, e^{-x} = \sum_{r=0}^\infty \frac{(-1)^r}{r!} \, (\nu)_r \, x^r,$$

$$\hat{\Gamma}_\nu = \frac{\Gamma(\nu + x\,\partial_x)}{\Gamma(\nu)}, \qquad (\nu)_r = \frac{\Gamma(\nu + r)}{\Gamma(\nu)}, \qquad \forall r \in \mathbb{R}_0^+.$$

$$(3.1.4)$$

The procedure of bringing the exponential outside the sign of integration, thus defining the operator $\hat{\Gamma}_\nu$, is allowed only for the values of $|x| < 1$ for which the series, containing the *Pochhammer symbol* $(\nu)_r$, converges. The series appearing in (3.1.4) is recognized as the Newton binomial, even though obtained in an *involved* albeit useful way for the purposes of the present intent.

In the spirit of umbral calculus we reinterpret the function $e^{(\nu)}(x)$ as an *ordinary exponential function*, by introducing the following definition.

Definition 11. The operator $\hat{\gamma}$ is defined as

$$\hat{\gamma}^r \nu_0 := (\nu)_r\,, \qquad (3.1.5)$$

with ν_0, the vacuum.

Accordingly, we can cast the function $e^{(\nu)}(x)$ in the form

$$e^{(\nu)}(x) = e^{-\hat{\gamma}\,x} \nu_0, \qquad (3.1.6)$$

thus formally treating it as an exponential (namely a transcendental function) even though the series (3.1.4) has a limited range of convergence $|x| < 1$.

As explained in what follows, we take advantage of the previous exponential umbral restyling of the function in (3.1.1) to construct a new formalism useful for the study of various familes of special polynomials.

Proposition 16. *By keeping the derivative of both sides of Eq. (3.1.6) with respect to the x variable and using the ordinary rules of calculus, we obtain $\forall x \in \mathbb{R}, \mathrm{Re}(\nu) > 0$, the identity*[a]

$$\partial_x e^{(\nu)}(x) = \left(\partial_x e^{-\hat{\gamma} x}\right) \nu_0 = -\hat{\gamma} e^{-\hat{\gamma} x} \nu_0 = \sum_{r=0}^{\infty} (-1)^{r+1} \hat{\gamma}^{r+1} \frac{x^r}{r!} \nu_0$$

$$= -\sum_{r=0}^{\infty} (-1)^r (\nu)_{r+1} \frac{x^r}{r!} = -\nu \, e^{(\nu+1)}(x), \tag{3.1.7}$$

which follows as a consequence of [3]

$$(\nu)_{r+1} = \nu \, (\nu + 1)_r \tag{3.1.8}$$

and more generally, $\forall m, r \in \mathbb{R}_0^+$.

$$(\nu)_{r+m} = (\nu)_m (\nu + m)_r. \tag{3.1.9}$$

We can accordingly state the rule

$$(\partial_x^m e^{-\hat{\gamma} x}) \nu_0 = (-1)^m (\hat{\gamma}^m e^{-\hat{\gamma} x}) \nu_0 = (-1)^m (\nu)_m e^{(\nu+m)}(x). \tag{3.1.10}$$

Before proceeding further it is worth clarifying a point, which will be more thoroughly treated later in Chapter 5. Even though the formalism we have developed allows to treat not trivial functions in terms of elementary exponential functions, some properties like the *semigroup identities* associated with the exponential case are not

---

[a]The same identity can be obtained from Eq. (3.1.2), which yields

$$\partial_x e^{(\nu)}(x) = -\frac{1}{\Gamma(\nu)} \int_0^{\infty} e^{-s} s^{\nu} e^{-sx} ds = -\frac{\Gamma(\nu+1)}{\Gamma(\nu)} e^{(\nu+1)}(x).$$

The umbral identity we have derived is not limited by any convergence restriction.

easily associated with $e^{(\nu)}(x)$. We underline indeed the observation as follows.

Observation 3. Albeit the following chain of identities is correct $\forall x, y \in \mathbb{R}$:

$$e^{(\nu)}(x+y) = e^{-\hat{\gamma}(x+y)}\nu_0 = e^{-\hat{\gamma}x}e^{-\hat{\gamma}y}\nu_0, \qquad (3.1.11)$$

it is also true that

$$e^{(\nu)}(x+y) \neq e^{(\nu)}(x)\,e^{(\nu)}(y). \qquad (3.1.12)$$

To overcome such an apparently paradoxical conclusion, we clarify that the concept of semigroup has to be properly framed within the *appropriate algebraic context*. In defining the semigroup and, thereby, the associated identities, the corresponding binary operations between x and y need to be defined.

The *associative binary operation (ABO)* $e^{x+y} = e^x e^y = e^y e^x$ is a consequence of the fact that $(x+y)^n = \sum_{r=0}^{n} \binom{r}{n} x^{n-r} y^r$. This means that we can define the opportune *ABO* if we modify the Newton binomial as follows.

Definition 2. Let $B(x, y)$ be the Beta function (1.0.5), we introduce a modified Newton binomial

$$(x \oplus_\nu y)^n := \sum_{r=0}^{n} \binom{n}{r} \left\{ \frac{(\nu)_n}{(\nu)_r} \right\}^{-1} x^{n-r} y^r,$$

$$\qquad (3.1.13)$$

$$\left\{ \frac{(\nu)_m}{(\nu)_p} \right\} = \frac{(\nu)_m}{(\nu)_{m-p}(\nu)_p} = \frac{B(\nu+m, \nu)}{B(\nu+m-p, \nu+p)}.$$

The corresponding *ABO* is

$$e^{(\nu)}(y)\,e^{(\nu)}(x) = \sum_{r=0}^{\infty} \frac{(\nu)_r}{r!}(-y)^r \sum_{s=0}^{\infty} \frac{(\nu)_s}{s!}(-x)^s$$

$$= \sum_{n=0}^{\infty} \frac{(-1)^n}{n!}(\nu)_n \left(\sum_{r=0}^{n} \binom{n}{r} \left\{ \frac{(\nu)_n}{(\nu)_r} \right\}^{-1} x^{n-r} y^r \right)$$

$$= e^{(\nu)}(x \oplus_\nu y). \qquad (3.1.14)$$

Accordingly, we conclude that the proper environment for the algebraic semi-group property of the umbral exponential discussed in this section is the use of associative operations of the type (3.1.14).

The reliability of the formalism we are developing can be further checked by deriving integrals involving the *pseudo Gaussian function* as shown in the following examples.

Example 31. If we have a function of the type

$$e^{(\nu)}(x^2) = e^{-\hat{\gamma} x^2} \nu_0 = (1 + x^2)^{-\nu}, \quad \forall x \in \mathbb{R}, \tag{3.1.15}$$

according to the rules we have stipulated along with the properties of the ordinary Gaussian function (1.2.2), we can state that[b] [99]

$$\int_{-\infty}^{+\infty} e^{(\nu)}(x^2)\, dx = \int_{-\infty}^{+\infty} e^{-\hat{\gamma} x^2}\, dx\, \nu_0 = \sqrt{\frac{\pi}{\hat{\gamma}}}\, \nu_0 = \sqrt{\pi}\, (\nu)_{-\frac{1}{2}}$$

$$= \sqrt{\pi}\, \frac{\Gamma\left(\nu - \frac{1}{2}\right)}{\Gamma\left(\nu\right)}, \qquad \mathrm{Re}(\nu) > \frac{1}{2}. \tag{3.1.16}$$

It must be stressed that the integral in Eq. (3.1.16) is extended to all the real axis and therefore the umbral representation should be representative of the function on the right-hand side of Eq. (3.1.15) and not of the relevant series expansion $\sum_{r=0}^{\infty}(-1)^r (\nu)_r \frac{x^{2r}}{r!}$, having radius of convergence $|x| < 1$. To clarify this point, we note that, by exploiting again the Laplace transform method and (1.2.2), we can alternatively write the integral (3.1.16) as

$$\int_{-\infty}^{+\infty} \frac{1}{(1 + x^2)^{\nu}}\, dx = \int_{-\infty}^{+\infty} dx \left[\frac{1}{\Gamma(\nu)} \int_0^{\infty} e^{-s} s^{\nu-1} e^{-s x^2}\, ds \right]$$

$$= \frac{\sqrt{\pi}}{\Gamma(\nu)} \int_0^{\infty} e^{-s} s^{\nu-\frac{3}{2}}\, ds = \sqrt{\pi}\, \frac{\Gamma\left(\nu - \frac{1}{2}\right)}{\Gamma\left(\nu\right)},$$

$$\tag{3.1.17}$$

which confirms the reliability of following the previously stated umbral rules.

---

[b]This (well-known) result is a byproduct of the outlined technique, but it could be also derived as a consequence of the (RMT) (see Section 7.6).

By pushing further the formalism we can take advantage of the wealth of properties of Gaussian integrals, by getting, e.g., the integral in Example 32.

Example 32. $\forall a, b, \nu \in \mathbb{R} : \mathrm{Re}(\nu) > \frac{1}{2}, \mathrm{Re}(a) > 0$

$$\int_{-\infty}^{+\infty} e^{(\nu)}\left(a\,x^2 + i\,b\,x\right) dx = \sqrt{\frac{\pi}{a\hat{\gamma}}}\, e^{-\hat{\gamma}\frac{b^2}{4a}}\, \nu_0$$

$$= \sqrt{\frac{\pi}{a}}\, \frac{\Gamma\left(\nu - \frac{1}{2}\right)}{\Gamma(\nu)}\, \frac{1}{\left(1 + \frac{b^2}{4a}\right)^{\nu - \frac{1}{2}}}. \qquad (3.1.18)$$

Let us now consider a further application of the previous procedure, by keeping the *successive derivatives* (with respect to the variable x) of the pseudo Gaussian function.

Example 33. We introduce, $\forall n \in \mathbb{N}, \ \forall x \in \mathbb{R}, \ \mathrm{Re}(\nu) > 0$,

$$e_n^{(\nu)}(x^2) := \partial_x^n e^{(\nu)}(x^2). \qquad (3.1.19)$$

We take advantage of the analogy with the properties of ordinary Gaussians, from the associated identity $\partial_x^n e^{a\,x^2} = H_n(2\,a\,x, a)\, e^{a\,x^2}$ in Eq. (1.2.7) and from (1.2.6), we adapt Eq. (3.1.6) to the pseudo-Gaussian case and we find

$$e_n^{(\nu)}(x^2) = H_n(-2\,\hat{\gamma}\,x, -\hat{\gamma})\, e^{-\hat{\gamma}\,x^2}\, \nu_0$$

$$= (-1)^n n! \sum_{r=0}^{\lfloor \frac{n}{2} \rfloor} \frac{(-1)^r (2\,x)^{n-2\,r}}{(n - 2\,r)!\, r!} \left(\hat{\gamma}^{n-r} e^{-\hat{\gamma}\,x^2}\right) \nu_0. \qquad (3.1.20)$$

On account of Eq. (3.1.10), we note that

$$\left(\hat{\gamma}^{n-r} e^{-\hat{\gamma}\,x^2}\right) \nu_0 = (\nu)_{n-r}\, e^{(\nu+n-r)}(x^2). \qquad (3.1.21)$$

If we now introduce the two-variable polynomials $\forall \xi, \eta \in \mathbb{R}, \ \forall n \in \mathbb{N}, \ \mathrm{Re}(\nu) > 0$,

$$K_n^{(\nu)}(\xi, \eta) := n! \sum_{r=0}^{\lfloor \frac{n}{2} \rfloor} \frac{(\nu)_{n-r}\, \xi^{n-2\,r}\eta^r}{(n - 2\,r)!\, r!}, \qquad (3.1.22)$$

we can recast Eq. (3.1.20) in the non-operatorial form

$$e_n^{(\nu)}(x^2) = (-1)^n K_n^{(\nu)}\left(\frac{2x}{1+x^2}, -\frac{1}{1+x^2}\right)\frac{1}{(1+x^2)^\nu}. \qquad (3.1.23)$$

For $\xi = 2x$, $y = -1$, the polynomials (3.1.22) reduce to the ordinary *Gegenbauer polynomials*, namely[c]

$$K_n^{(\nu)}(2x, -1) = n!\,C_n^{(\nu)}(x). \qquad (3.1.24)$$

Furthermore, the identity

$$(-1)^n\,e^{(-\nu)}(x^2)\,e_n^{(\nu)}(x^2) = K_n^{(\nu)}\left(\frac{2x}{1+x^2}, -\frac{1}{1+x^2}\right) \qquad (3.1.25)$$

can be viewed as the associated Rodriguez formula [3].

It is also worth stressing that the discussed formalism and the use of the relation $\partial_x^n e^{a\,x^2 + b\,x} = H_n(2\,a\,x + b,\,a)\,e^{a\,x^2 + b\,x}$ in Eq. (2.5.9) yield the result

$$e_n^{(\nu)}(a\,x^2 + b\,x)$$
$$= (-1)^n K_n^{(\nu)}\left(\frac{2\,a\,x + b}{1 + a\,x^2 + b\,x}, -\frac{a}{1 + a\,x^2 + b\,x}\right)\frac{1}{(1 + a\,x^2 + b\,x)^\nu}. \qquad (3.1.26)$$

The results we have presented in this introduction disclose one of the advantages of formalism, which allows the *derivation of the properties of Gegenbauer polynomials from those of Hermite*. Further consequences of this point of view will be discussed in the following sections.

3.1.1 *Gegenbauer Polynomials*

The study of the properties of the Gegenbauer polynomials is based on two-variable Hermite polynomials of order 2. It is natural to

---

[c]We call the definition of Gegenbauer polynomials given in [2]

$$C_n^{(\nu)}(x) = \sum_{k=0}^{\lfloor\frac{n}{2}\rfloor}(-1)^k\frac{\Gamma(n - k + \nu)}{\Gamma(\nu)k!(n - 2k)!}(2x)^{n-2k}.$$

conclude that higher-order *HP* are tailor-made to define generalized forms of Gegenbauer polynomials.

Proposition 17. *Let* $\forall x, y \in \mathbb{R}$, $\forall n, m \in \mathbb{N}$,

$$H_n^{(m)}(x,y) = n! \sum_{r=0}^{\lfloor \frac{n}{m} \rfloor} \frac{x^{n-m\,r} y^r}{(n-m\,r)!\,r!},$$

$$\sum_{n=0}^{\infty} \frac{t^n}{n!} H_n^{(m)}(x,y) = e^{xt+yt^m},$$

(3.1.27)

be the higher-order two-variable Hermite polynomials (see Section 7.4) [5], then, according to our formalism, we can identify

$$K_n^{(\nu,\,m)}(\xi,\,\eta) = H_n^{(m)}(\hat\gamma\,\xi,\,\hat\gamma\,\eta)\,\nu_0 \tag{3.1.28}$$

and obtain, $\forall t \in \mathbb{R} :\mid t \mid < 1,$

$$\sum_{n=0}^{\infty} \frac{t^n}{n!} K_n^{(\nu,\,m)}(-\xi,\,-\eta) = e^{(\nu)}(\xi\,t + \eta\,t^m) = \frac{1}{(1+\xi\,t+\eta\,t^m)^\nu}.$$

(3.1.29)

The repeated derivatives of functions like $e^{(\nu)}(ax^m + bx)$ *can be expressed by using the properties of the higher-order Hermite Kampé de Fériét polynomials and of their generalized forms. The use of the following identity involving multivariable HP (see [18]):*

$$\partial_x^n e^{P(x)} = H_n^{(m,m-1,\dots,2)}\left(P'(x), \frac{P''(x)}{2}, \frac{P'''(x)}{3!}, \dots, \frac{P^{(m)}(x)}{m!}\right) e^{P(x)},$$

$$P(x) = a\,x^m + b\,x,$$

(3.1.30)

can be exploited, along with Eq. (3.1.30), to get (see also Eq. (3.1.20))

$$\partial_x^n e^{(\nu)}(a\,x^m + b\,x) = e_n^{(\nu)}(a\,x^m + b\,x)$$

$$= H_n^{(m,\,m-1,\dots,2)}\left(-\hat\gamma\,P'(x),\, -\hat\gamma\frac{P''(x)}{2},\, -\hat\gamma\frac{P'''(x)}{3!}, \dots, -\hat\gamma\frac{P^{(m)}(x)}{m!}\right)$$

$$\cdot\, e^{-\hat\gamma\,P(x)} \nu_0$$

$$= K_n^{(\nu, m, m-1,\ldots,2)} \left(-\frac{P'(x)}{P(x)+1}, -\frac{P''(x)}{2(P(x)+1)}, -\frac{P'''(x)}{3!(P(x)+1)}, \ldots, \right.$$

$$\left. -\frac{P^{(m)}(x)}{m!(P(x)+1)} \right) \frac{1}{(P(x)+1)^\nu}, \tag{3.1.31}$$

where

$$K_n^{(\nu, m, m-1,\ldots,2)}(x_1, x_2, x_3, \ldots, x_m)$$

$$= \frac{1}{\Gamma(\nu)} \int_0^\infty e^{-s} s^{\nu-1} H_n^{(m, m-1,\ldots,2)}(x_1 s, x_2 s, x_3 s, \ldots, x_m s) ds,$$

$$H_n^{(m, m-1,\ldots,2)}(x_1, x_2, x_3, \ldots, x_m) = n! \sum_{r=0}^{\lfloor \frac{n}{m} \rfloor} \frac{x_m^r H_{n-mr}^{(m-1,\ldots,2)}(x_1, \ldots, x_{m-1})}{(n-mr)! r!}$$

$$\sum_{n=0}^\infty \frac{t^n}{n!} H_n^{(\{m\})}(\{x\}) = e^{\sum_{s=1}^m x_s t^s},$$

$$\{m\} = m, m-1, \ldots, 2; \quad \{x\} = x_1, x_2, \ldots, x_m. \tag{3.1.32}$$

The same method allows some progress in the derivation of Gegenbauer generating functions *and indeed, by exploiting Eq.* (2.4.15), *we find*

$$\sum_{n=0}^\infty \frac{t^n}{n!} K_{n+l}^{(\nu)}(-\xi, -\eta) = H_l(-\hat{\gamma}(\xi + 2\eta t), -\hat{\gamma}\eta) e^{-\hat{\gamma}(\xi t + \eta t^2)} \nu_0$$

$$= \frac{(-1)^l K_l^{(\nu)} \left(\frac{(\xi + 2\eta t)}{1 + \xi t + \eta t^2}, -\frac{\eta}{1 + \xi t + \eta t^2} \right)}{(1 + \xi t + \eta t^2)^\nu}, \quad |t| < \left| \frac{\xi - \sqrt{\xi^2 - 4\eta}}{2\eta} \right|, \tag{3.1.33}$$

which can be easily derived from the corresponding case of the Hermite polynomials $\sum_{n=0}^\infty \frac{t^n}{n!} H_{n+l}(x, y) = H_l(x+2yt, y) e^{xt+yt^2}$ [18].

All the previous results can be obtained by the use of the Laplace transform method. The integral transforms are indeed not an alternative, but the rigorous support of the umbral methods we are developing.

3.1.2 *Jacobi Polynomials*

In the previous section, we have exploited a, likely, powerful tool to deal with a plethora of problems concerning the theory of special functions, whose relevant technicalities can accordingly be reduced to straightforward exercises in elementary calculus.

Let us now take a step further, by introducing the following polynomials.

Definition 13. We define a new family of two-variable polynomials $\forall \xi, \eta, \alpha, \beta \in \mathbb{R}, \ \forall n \in \mathbb{N}$

$$\frac{1}{n!} R_n^{(\alpha, \beta)}(\xi, \eta) := \hat{c}_1^\alpha \hat{c}_2^\beta \left[\hat{c}_1 \xi + \hat{c}_2 \eta \right]^n \varphi_{1,0} \varphi_{2,0}, \qquad (3.1.34)$$

where the operators $\hat{c}$ labeled by two different indexes act on two different vacua as

$$\hat{c}_1^\nu \hat{c}_2^\mu \varphi_{1,0} \varphi_{2,0} = (\hat{c}_1^\nu \varphi_{1,0})(\hat{c}_2^\mu \varphi_{2,0}) = \frac{1}{\Gamma(\nu+1)} \cdot \frac{1}{\Gamma(\mu+1)}. \qquad (3.1.35)$$

Corollary 13. *According to the previous definition, we obtain the following explicit form for the polynomials defined in Eq.* (3.1.34):

$$R_n^{(\alpha, \beta)}(\xi, \eta) = (n!)^2 \sum_{s=0}^n \frac{\xi^{n-s} \eta^s}{[(n-s)!] \, s! \, \Gamma(n-s+\alpha+1) \Gamma(s+\beta+1)}. \tag{3.1.36}$$

The relevant properties can easily be derived by the use of elementary algebraic manipulations.

Properties 7.

$$\frac{1}{(n+1)!} R_{n+1}^{(\alpha, \beta)}(\xi, \eta) = [\hat{c}_1 \xi + \hat{c}_2 \eta] \, \hat{c}_1^\alpha \hat{c}_2^\beta \left[\hat{c}_1 \xi + \hat{c}_2 \eta \right]^n \varphi_{1,0} \varphi_{2,0}$$

$$= \frac{1}{n!} \left(\xi \, R_n^{(\alpha+1, \beta)}(\xi, \eta) + \eta \, R_n^{(\alpha, \beta+1)}(\xi, \eta) \right) \tag{3.1.37}$$

and that

$$
\begin{aligned}
\partial_\xi R_n^{(\alpha,\,\beta)}(\xi,\,\eta) &= n^2\, R_{n-1}^{(\alpha+1,\,\beta)}(\xi,\,\eta), \\
\partial_\eta R_n^{(\alpha,\,\beta)}(\xi,\,\eta) &= n^2\, R_{n-1}^{(\alpha,\,\beta+1)}(\xi,\,\eta).
\end{aligned}
\tag{3.1.38}
$$

Furthermore, we can determine its generating functions by the use of analogous elementary procedures.

Properties 8.

$$
\sum_{n=0}^{\infty} \frac{t^n}{(n!)^2} R_n^{(\alpha,\,\beta)}(\xi,\,\eta) = \hat{c}_1^\alpha \hat{c}_2^\beta\, e^{\,t\,(\hat{c}_1\xi + \hat{c}_2\eta)} \varphi_{1,\,0}\varphi_{2,\,0}, \quad \forall t \in \mathbb{R} \tag{3.1.39}
$$

and, if we note that, by generalizing Tricomi–Bessel function (2.5.16) $\forall \nu \in \mathbb{R} : Re(\nu) > 0$, we get

$$
\hat{c}^\nu e^{-\hat{c}x}\varphi_0 = \sum_{r=0}^{\infty} \frac{(-x)^r \hat{c}^{r+\nu}}{r!}\varphi_0 = C_\nu(x),
$$
$$
C_\nu(x) = \sum_{r=0}^{\infty} \frac{(-1)^r x^r}{r!\,\Gamma(r+\nu+1)},
\tag{3.1.40}
$$

which are linked to the cylindrical Bessel functions by (see Eq. (1.2.18) in generalized form)

$$
C_\nu(x) = \left(\frac{1}{x}\right)^{\frac{\nu}{2}} J_\nu(2\sqrt{x}),
\tag{3.1.41}
$$

we can write the generating function (3.1.39) in terms of a product of Bessel functions[d] $\forall t \in \mathbb{R}$

$$
\sum_{n=0}^{\infty} \frac{t^n}{(n!)^2} R_n^{(\alpha,\,\beta)}(\xi,\,\eta) = C_\alpha(-\xi\, t)\, C_\beta(-\eta\, t)
$$

$$
= \frac{1}{\sqrt{(\xi^\alpha \eta^\beta)\, t^{\alpha+\beta}}} I_\alpha(2\sqrt{\xi\, t}) I_\beta\left(2\sqrt{\eta\, t}\right).
\tag{3.1.42}
$$

---

[d]Where $I_\nu(x) = (-i)^\nu J_\nu(ix)$ is the first kind of modified Bessel function (see Chapter 4).

Definition 14. The polynomials $R_n^{(\alpha, \beta)}(\xi, \eta)$, $\forall \xi, \eta, \alpha, \beta \in \mathbb{R}, \forall n \in \mathbb{N}$, can be used to define the ordinary *Jacobi polynomials* $\forall x \in \mathbb{R}$ [3] through the identity

$$P_n^{(\alpha, \beta)}(x) = \frac{\Gamma(n + \alpha + 1)\Gamma(n + \beta + 1)}{(n!)^2} R_n^{(\alpha, \beta)}(\xi(x), \eta(x)),$$

$$\xi(x) := \frac{x - 1}{2}, \qquad \eta(x) := \frac{1 + x}{2}.$$

(3.1.43)

Properties 9. The relevant recurrences are obtained from Eqs. (3.1.37)–(3.1.38) and we can write

$$(n + 1)P_{n+1}^{(\alpha, \beta)}(x)$$

$$= \frac{1}{2}x\left[(n + \beta + 1)\,P_n^{(\alpha+1, \beta)}(x) + (n + \alpha + 1)\,P_n^{(\alpha, \beta+1)}(x)\right]$$

$$- \frac{1}{2}\left[(n + \beta + 1)\,P_n^{(\alpha+1, \beta)}(x) - (n + \alpha + 1)\,P_n^{(\alpha, \beta+1)}(x)\right]$$

(3.1.44)

and

$$\frac{d}{dx}P_n^{(\alpha, \beta)}(x) = \frac{\Gamma(n - \alpha + 1)\Gamma(n + \beta + 1)}{2(n - 1)!^2}$$

$$\cdot \left(R_{n-1}^{(\alpha+1, \beta)}(\xi(x), \eta(x)) + R_{n-1}^{(\alpha, \beta+1)}(\xi(x), \eta(x))\right)$$

$$= \frac{1}{2}\left[(n + \beta)P_{n-1}^{(\alpha+1, \beta)}(x) + (n + \alpha)P_{n-1}^{(\alpha, \beta+1)}(x)\right]$$

$$= \frac{n + \alpha + \beta + 1}{2}P_{n-1}^{(\alpha+1, \beta+1)}(x).$$

(3.1.45)

Properties 10. The relevant generating function can be written as

$$\sum_{n=0}^{\infty} \frac{t^n}{\Gamma(n + \alpha + 1)\Gamma(n + \beta + 1)}P_n^{(\alpha, \beta)}(x)$$

$$= \left(\frac{2}{\sqrt{2(x - 1)t}}\right)^{\alpha}\left(\frac{2}{\sqrt{2(x + 1)t}}\right)^{\beta} I_\alpha(\sqrt{2(x - 1)t})I_\beta\left(\sqrt{2(x + 1)t}\right).$$

(3.1.46)

We can now deduce further consequences from the previous umbral restyling of the Jacobi polynomials.

Proposition 18. *The index doubling "Theorem" can be derived by noting that*

$$R_{2n}^{(\alpha,\,\beta)}(\xi,\,\eta) = (2n)!\,\hat{c}_1^{\alpha}\hat{c}_2^{\beta}\,[\hat{c}_1\xi + \hat{c}_2\eta]^n\,[\hat{c}_1\xi + \hat{c}_2\eta]^n\,\varphi_{1,\,0}\varphi_{2,\,0}$$

$$= \frac{(2n)!}{n!}\sum_{s=0}^{n}\binom{n}{s}\xi^{n-s}\eta^{s}R_n^{(n-s+\alpha,\,s+\beta)}(\xi,\,\eta) \quad (3.1.47)$$

which, on account of Eq. (3.1.43), yields

$$P_{2n}^{(\alpha,\,\beta)}(x)$$

$$= \frac{n!}{(2n)!}\Gamma(2n+\alpha+1)\,\Gamma(2n+\beta+1)\sum_{s=0}^{n}\binom{n}{s}\,(\xi(x))^{n-s}\,(\eta(x))^{s}\,\cdot$$

$$\cdot\frac{P_n^{(n-s+\alpha,\,s+\beta)}(x)}{\Gamma(2n-s+\alpha+1)\,\Gamma(n+s+\beta+1)}.$$

$$(3.1.48)$$

Furthermore, an analogous procedure yields the following proposition.

Proposition 19. *The argument duplication formula is*

$$P_n^{(\alpha,\,\beta)}(2x) = \frac{\Gamma(n+\alpha+1)\,\Gamma(n+\beta+1)}{n!}\sum_{s=0}^{n}\binom{n}{s}\left(\frac{x}{2}\right)^{s}\sum_{r=0}^{s}\binom{s}{r}$$

$$\cdot\frac{(n-s)!P_{n-s}^{(s-r+\alpha,\,r+\beta)}(x)}{\Gamma(n-r+\alpha+1)\Gamma(n-s+r+\beta+1)}.$$

$$(3.1.49)$$

It is evident that the method is so straightforward that all the previous identities can easily be generalized, as touched on in the following.

Proposition 20. *The associated Laguerre polynomials* (2.7.6)– (2.7.7) $\left(L_n^{(\alpha)}(x,\,y) - \Lambda_n^{(\alpha)}(x,\,y)\right)$ *further confirm the mutual link with the Jacobi family and we find indeed that*

$$R_n^{(\alpha,\beta)}(\xi,\eta) = (n!)^2 \sum_{s=0}^{n} \frac{(-1)^s L_{n-s}^{(\alpha)}(-\xi,\xi+\eta) L_s^{(\beta)}(\eta,\xi+\eta)}{\Gamma(n-s+\alpha+1)\Gamma(s+\beta+1)},$$

$$P_n^{(\alpha,\beta)}(x) = \frac{\Gamma(n+\alpha+1)\Gamma(n+\beta+1)}{n!} \hat{c}_1^\alpha \hat{c}_2^\beta$$

$$\cdot \left[\left(y + \hat{c}_1 \frac{x-1}{2} \right) - \left(y - \hat{c}_2 \frac{x+1}{2} \right) \right]^n \varphi_{1,0}\varphi_{2,0}$$

$$= \Gamma(n+\alpha+1)\Gamma(n+\beta+1) \cdot \sum_{s=0}^{n} \frac{(-1)^s L_{n-s}^{(\alpha)}\left(\frac{1-x}{2}, y\right) L_s^{(\beta)}\left(\frac{x+1}{2}, y\right)}{\Gamma(n-s+\alpha+1)\Gamma(s+\beta+1)}.$$

$$(3.1.50)$$

We have covered some of the properties of Jacobi polynomials by employing a minimal computational effort, we have fixed the formalism we are going to use and have provided an idea of the consequences which can be drawn by means of these methods.

3.1.3   *Legendre Polynomials*

The Legendre polynomials are a particular case of the Jacobi family [3] and can be identified as follows.

Corollary 14. *According to the positions in (3.1.43) $\forall x \in \mathbb{R}, \forall n \in \mathbb{N}$, we get*

$$P_n(x) = P_n^{(0,0)}(x) = R_n^{(0,0)}(\xi,\eta). \qquad (3.1.51)$$

Their properties can be therefore derived as a consequence of those of the R_n polynomials in the particular case of $\alpha = \beta = 0$. Let us therefore go back to Eq. (3.1.50) and note that

Properties 11.

$$\frac{1}{n!}P_n(x) = \left[\hat{c}_1 \frac{x-1}{2} + \hat{c}_2 \frac{x+1}{2} \right]^n \varphi_{1,0}\varphi_{2,0},$$

$$\frac{1}{n!}P_n(0) = \left(-\frac{\hat{c}_1}{2} + \frac{\hat{c}_2}{2} \right)^n \varphi_{1,0}\varphi_{2,0} = \frac{1}{n!}R_n\left(-\frac{1}{2}, \frac{1}{2} \right)$$

$$= \frac{(-1)^n}{2^n n!} \sum_{s=0}^{n} (-1)^s \left(\frac{n!}{s!\,(n-s)!} \right)^2,$$

$$P_n(1) = R_n(0,\, 1) = 1, \quad P_n(-1) = R_n(-1,\, 0) = (-1)^n.$$

$$(3.1.52)$$

The use of the auxiliary polynomials R_n is a fairly important tool to state further identities, as, e.g.,

$$P_n(\lambda x) = n! \left[\hat{c}_1 \frac{\lambda x - 1}{2} + \hat{c}_2 \frac{\lambda x + 1}{2} \right]^n \varphi_{1,0} \varphi_{2,0}$$

$$= n! \left[\lambda \left(\hat{c}_1 \frac{x-1}{2} + \hat{c}_2 \frac{x+1}{2} \right) + \hat{c}_1 \frac{\lambda - 1}{2} + \hat{c}_2 \frac{-\lambda+1}{2} \right]^n \varphi_{1,0} \varphi_{2,0}$$

$$= (n!)^2 \sum_{s=0}^{n} \lambda^{n-s} \sum_{r=0}^{s} \frac{\xi(\lambda)^{s-r} \eta(-\lambda)^r P_{n-s}^{(s-r,\, r)}(x)}{(s-r)!\, r!\, (n-r)!\, (n-s+r)!}. \qquad (3.1.53)$$

Furthermore, we obtain

$$P_{n+m}(x) = (n+m)! \left[\hat{c}_1 \frac{x-1}{2} + \hat{c}_2 \frac{x+1}{2} \right]^{n+m} \varphi_{1,0} \varphi_{2,0}$$

$$= n!\, m! \sum_{s=0}^{m} \binom{n+m}{s} \frac{\xi(x)^{m-s} \eta(x)^s P_n^{(m-s,\, s)}(x)}{(m-s)!\, (n+s)!} \qquad (3.1.54)$$

and

$$P_n(x+y) = (n!)^2 \sum_{s=0}^{n} \left(\frac{y}{2} \right)^s \sum_{r=0}^{s} \frac{P_{n-s}^{(s-r,\, r)}(x)}{(s-r)!\, r!\, (n-r)!\, (n-s+r)!}. \qquad (3.1.55)$$

The previous identity cannot be considered an "addition Theorem" in the strict sense, but rather a Taylor series expansion.

The next step is the derivation of the differential equation satisfied by the Legendre polynomials.

Proposition 21. *By the use of Eqs.* (3.1.44), (3.1.45), *we get*

$$n\,P_{n-1}(x) = \left[(1-x^2)\frac{d}{dx} + n\,x\right] P_n(x),$$

$$(n+1)P_{n+1}(x) = \left\{(2n+1)\,x - \left[(1-x^2)\frac{d}{dx} + n\,x\right]\right\} P_n(x).$$

$$(3.1.56)$$

By combining the previous recurrences, we can introduce the following operators:

$$\hat{N}_- = (1-x^2)\frac{d}{dx} + \hat{n}\,x, \quad \hat{N}_+ = -(1-x^2)\frac{d}{dx} + (\hat{n}+1)\,x, \quad (3.1.57)$$

defined in such a way that

$$\hat{N}_- P_n(x) = n\,P_{n-1}(x), \quad \hat{N}_+ P_n(x) = (n+1)\,P_{n+1}(x), \quad (3.1.58)$$

where $\hat{n}$ is a kind of number operator "counting" the index of the Legendre polynomial, namely

$$\hat{n}\,P_{m+k}(x) = (m+k)\,P_{m+k}(x). \quad (3.1.59)$$

According to the previous definitions, we find

$$\hat{N}_+\hat{N}_- P_n(x) = \left[-(1-x^2)\frac{d}{dx} + (\hat{n}+1)x\right]\left[(1-x^2)\frac{d}{dx} + \hat{n}x\right] P_n(x)$$

$$= \left[-(1-x^2)\frac{d}{dx} + n\,x\right]\left[(1-x^2)\frac{d}{dx} + n\,x\right]$$

$$P_n(x) = n^2 P_n(x), \quad (3.1.60)$$

which explicitly yields the following second-order equation satisfied by the Legendre polynomials written in the form

$$\left(\frac{d}{dx}(1-x^2)\frac{d}{dx}\right) P_n(x) + n\,(n+1)\,P_n(x) = 0. \quad (3.1.61)$$

In the forthcoming section, we extend the umbral formalism to make further progress by including the properties of the associated Legendre polynomials.

3.1.4 Generalized Forms

In this section, we explore generalized forms of the presented polynomial families.

Proposition 22. *Let us consider the evaluation of the following repeated derivatives* $\forall m \in \mathbb{N}, \forall \alpha, \beta, a, b \in \mathbb{R}, \in Re(\nu) > 0, \forall x \in \mathbb{R} : Q(x) > -1:$

$$F_m^{(\nu)}(x) = \left(\frac{d}{dx}\right)^m \left(\frac{e^{P(x)}}{(1+Q(x))^\nu}\right),$$ (3.1.62)

$$P(x) = \alpha x^2 + \beta x, \qquad Q(x) = ax^2 + bx.$$

The use of the umbral procedure (3.1.6) allows a significant simplification of the relevant algebra. By setting indeed

$$\frac{e^{P(x)}}{(1+Q(x))^\nu} = e^{P(x)} e^{(\nu)}(Q(x)) = e^{P(x)-\hat{\gamma}Q(x)}\nu_0$$

$$= e^{(\alpha-\hat{\gamma}a)x^2+(\beta-\hat{\gamma}b)x}\nu_0,$$ (3.1.63)

we find, applying Eq. (2.5.9),

$$F_m^{(\nu)}(x) = H_m\left(2(\alpha - \hat{\gamma}a)x + (\beta - \hat{\gamma}b), \alpha - \hat{\gamma}a\right) e^{P(x)-\hat{\gamma}Q(x)}\nu_0.$$ (3.1.64)

The use of the so far developed rules yields

$$F_m^{(\nu)}(x) = m! \sum_{r=0}^{\lfloor \frac{m}{2} \rfloor} \frac{1}{(m-2r)!r!} \sum_{s=0}^{m-2r} \binom{m-2r}{s} A^{m-2r-s} B^s$$

$$\cdot \sum_{q=0}^{r} \binom{r}{q} (-1)^{s+q} \alpha^{r-q} a^q (\nu)_{s+q} \, e^{(\nu+s+q)}(Q(x)) e^{P(x)}$$

$$A = 2\alpha x + \beta; \qquad B = 2ax + b$$ (3.1.65)

or, in a more compact form,

$$F_m^{(\nu)}(x) = \Omega_m^{(\nu)}\left(P'(x), \frac{P''(x)}{2}; \frac{Q'(x)}{(1+Q(x))}, \frac{Q''(x)}{2(1+Q(x))}\right)$$

$$\cdot \frac{e^{P(x)}}{(1+Q(x))^{\nu}}$$

$$\Omega_m^{(\nu)}(x, y; u, z) = \sum_{s=0}^{m} \binom{m}{s}(-1)^s H_{m-s}(x, y) K_s^{(\nu)}(u, -z).$$

$$(3.1.66)$$

It is now worth to explore more accurately the role of the $K_n^{(\nu)}(.,.)$ polynomials introduced in Section 3.1. With this aim, we consider the following particular case.

Example 34. Let $\nu = \frac{1}{2}$, by using Eq. (3.1.23), we get

$$e_n^{(\frac{1}{2})}(x^2) = (-1)^n K_n^{(\frac{1}{2})}\left(\frac{2x}{1+x^2}, -\frac{1}{1+x^2}\right)\frac{1}{(1+x^2)^{\frac{1}{2}}}, \qquad (3.1.67)$$

furthermore, by recalling the identity (3.1.32),

$$K_n^{(\frac{1}{2})}(a, b) = \frac{1}{\sqrt{\pi}}\int_0^{\infty} e^{-s}s^{-\frac{1}{2}}H_n(as, bs)ds \qquad (3.1.68)$$

and, recalling Eq. (1.2.9) $H_n(x, y) = y^{\frac{n}{2}}H_n\left(\frac{x}{\sqrt{y}}, 1\right)$, we can easily infer that $\forall i \in \mathbb{R} : 1 - at - bt^2 > 0$,

$$\sum_{n=0}^{\infty}\frac{t^n}{n!}K_n^{(\frac{1}{2})}(a, b) = \frac{1}{\sqrt{1-at-bt^2}}, \qquad (3.1.69)$$

which is the *generating function of Legendre polynomials* for $a = 2x$, $b = -1$. Moreover, the use of the identity (1.2.9) yields

$$(1+x^2)^{\frac{n+1}{2}}e_n^{(\frac{1}{2})}(x^2) = (-1)^n n! P_n\left(\frac{x}{\sqrt{1+x^2}}\right), \qquad (3.1.70)$$

which can be extended to the cases involving the generalized Legendre forms.

The same procedure can be applied to derive the following generating function for ordinary Legendre.

Example 35. $\forall n, l \in \mathbb{N}, \forall x \in \mathbb{R}, \forall t \in \mathbb{R} : (1 - 2xt + t^2)^{l+1} > 0$, we find

$$\sum_{n=0}^{\infty} \binom{n+l}{l} t^n P_{n+l}(x) = \frac{P_l\left(\dfrac{x-t}{\sqrt{1-2xt+t^2}}\right)}{(1-2xt+t^2)^{\frac{l+1}{2}}}. \tag{3.1.71}$$

It is also a particular case of Eq. (3.1.33).

Corollary 15. *According to the above point of view, the mth derivative of the $P_n(x)$ can therefore be easily calculated, thus finding*

$$\left(\frac{d}{dx}\right)^m P_n(x) = \frac{2^m}{\sqrt{\pi}(n-m)!} \int_0^\infty e^{-s} s^{m-\frac{1}{2}} H_{n-m}(2xs, -s)ds$$

$$= \frac{1}{\sqrt{\pi}} \sum_{r=0}^{\lfloor \frac{n-m}{2} \rfloor} \frac{(-1)^r 2^{n-2r} x^{n-m-2r} \Gamma(n-r+\frac{1}{2})}{(n-m-2r)!r!}. \tag{3.1.72}$$

On the other hand, the successive derivatives of the Legendre polynomials can be obtained from (3.1.52) and yield the following link with the Jacobi polynomials:

$$\left(\frac{d}{dx}\right)^m P_n(x)$$

$$= \frac{1}{2^m} \frac{(n!)^2}{(n-m)!} (\hat{c}_1 + \hat{c}_2)^m \left[\hat{c}_1 \frac{x-1}{2} + \hat{c}_2 \frac{x+1}{2}\right]^{n-m} \varphi_{1,0}\varphi_{2,0}$$

$$= \frac{(n!)^2}{2^m} \sum_{s=0}^{m} \binom{m}{s} \frac{P_{n-m}^{(m-s,\,s)}(x)}{(n-s)!(n-m+s)!}. \tag{3.1.73}$$

We now discuss a further alternative formulation of the theory of Legendre polynomials, using a formalism touched upon in [62], which will be embedded with the technique developed until now.

Observation 4. We note that we can obtain a similar result at (2.3.2) by introducing a new family of polynomials $\forall x, y \in \mathbb{R}$, $\forall n \in \mathbb{N}$

$$\Pi_n(x, y) = \left((x + {}_y\hat{h}_c)^n \theta_0 \right) \varphi_0, \tag{3.1.74}$$

where (by using the $\hat{c}$-operator (1.1.6))

$$\left({}_y\hat{h}_c^r \, \theta_0 \right) \varphi_0 = \frac{y^{\frac{r}{2}} \, r!}{\Gamma\left(\frac{r}{2} + 1\right)} \left| \cos\left(r \frac{\pi}{2} \right) \right| \hat{c}^{\frac{r}{2}} \varphi_0 = \frac{y^{\frac{r}{2}} \, r!}{\Gamma\left(\frac{r}{2} + 1\right)^2} \left| \cos\left(r \frac{\pi}{2} \right) \right|. \tag{3.1.75}$$

According to the above definition (by following the proof of Proposition 8), we obtain the explicit expression for the Π polynomials as

$$\Pi_n(x, y) = H_n(x, \hat{c}\, y)\, \varphi_0 = n! \sum_{k=0}^{\lfloor \frac{n}{2} \rfloor} \frac{x^{n-2k} y^k}{(k!)^2 \, (n - 2k)!}. \tag{3.1.76}$$

They are essentially *hybrid Laguerre–Hermite polynomials* (including the operator ${}_y\hat{h}$ di Laguerre and the operator $\hat{c}$ di Hermite) satisfying the "heat" equation

$$\begin{cases} {}_l\partial_y G(x, y) = -\partial_x^2 G(x, y) \\ G(x, 0) = x^n, \end{cases} \tag{3.1.77}$$

with ${}_l\partial_y$ l-derivative (2.6.8). We note that the polynomials (3.1.76) can be defined through the operational rule (by using (1.1.20))

$$\Pi_n(x, y) = C_0(-y\, \partial_x^2)\, x^n \tag{3.1.78}$$

or also (by applying (1.2.18))

$$\Pi_n(x, y) = J_0 \left(2i\sqrt{y}\, \partial_x \right) x^n \tag{3.1.79}$$

and the Legendre polynomials can be identified with the particular case

$$P_n(x) = \Pi_n \left(x, -\frac{1 - x^2}{4} \right). \tag{3.1.80}$$

The procedure suggests that the *negative-order hybrid functions* can be written as

$$\Pi_{-\nu}(x,y) = \left((x + y\hat{h}_c)^{-\nu}\theta_0\right)\varphi_0$$

$$= \frac{1}{\Gamma(\nu)} \int_0^\infty s^{\nu-1}e^{-sx} J_0\left(2is\sqrt{y}\right) ds. \quad (3.1.81)$$

Infinite integrals containing, e.g., products of Bessel and exponential functions, like (2.7.5), can therefore be expressed in terms of negative-order Π functions. We find, for example, that

$$S(a,b) = \int_0^\infty e^{-ax} J_0(bx)dx = \Pi_{-1}\left(a, -\frac{b^2}{4}\right). \quad (3.1.82)$$

Finally, from the previous identities we find the *Legendre generating function*, $\forall t \in \mathbb{R}$,

$$\sum_{n=0}^\infty \frac{t^n}{n!} P_n(x) = e^{x\,t - \hat{c}\frac{1-x^2}{4}t^2}\varphi_0 = e^{x\,t} J_0\left[t\sqrt{1-x^2}\right]. \quad (3.1.83)$$

We can now derive a further consequence from the above equation and from the umbral definition of the Legendre polynomials (3.1.52), according to which we find

$$\sum_{n=0}^\infty \frac{t^n}{n!} P_n(x) = \sum_{n=0}^\infty t^n \left[\hat{c}_1 \frac{x-1}{2} + \hat{c}_2 \frac{x+1}{2}\right]^n \varphi_{1,0}\,\varphi_{2,0}$$

$$= \frac{1}{1 - t\left[\hat{c}_1\frac{x-1}{2} + \hat{c}_2\frac{x+1}{2}\right]}\varphi_{1,0}\,\varphi_{2,0}. \quad (3.1.84)$$

The use of standard Laplace transform identities yields

$$\frac{1}{1 - t\left[\hat{c}_1\frac{x-1}{2} + \hat{c}_2\frac{x+1}{2}\right]}\left[\varphi_{1,0}\,\varphi_{2,0}\right]$$

$$= \int_0^\infty e^{-s}e^{s\,t\left(\hat{c}_1\frac{x-1}{2}+\hat{c}_2\frac{x+1}{2}\right)}ds\,\varphi_{1,0}\,\varphi_{2,0}$$

$$= \int_0^\infty e^{-s}C_0\left(\frac{1-x}{2}s\,t\right) C_0\left(-\frac{1+x}{2}s\,t\right) ds, \quad (3.1.85)$$

which once confronted with (3.1.83) yields

$$\int_0^\infty e^{-s} C_0\left(\frac{1-x}{2}st\right) C_0\left(-\frac{1+x}{2}st\right) ds = e^{xt} J_0\left(t\sqrt{1-x^2}\right).$$

$$(3.1.86)$$

Before concluding this part, for the proposed umbral definition of the Gegenbauer polynomials, we can adopt an analogous point of view by noting that the use of the operational identity

$$M(a\,x,\,b\,y) = a^{x\,\partial_x} b^{y\,\partial_y} M(x,\,y), \qquad (3.1.87)$$

based on the Euler dilation operator [122], allows the following reshuffling of Eq. (3.1.32)

$$K_n^{(\nu)}(\xi,\,\eta) = \frac{1}{\Gamma(\nu)} \int_0^\infty e^{-s} s^{\nu-1} s^{\xi\,\partial_\xi+\eta\,\partial_\eta} ds\, H_n(\xi,\,\eta) \qquad (3.1.88)$$

and the use of the properties of the Gamma function eventually leads to the following operational definition of the Gegenbauer polynomials

$$K_n^{(\nu)}(\xi,\,\eta) = \hat\Gamma_\nu\, H_n(\xi,\,\eta), \qquad \hat\Gamma_\nu = \frac{\Gamma(\nu+\xi\,\partial_\xi+\eta\,\partial_\eta)}{\Gamma(\nu)}. \qquad (3.1.89)$$

The operator $\hat\Gamma_\nu$ is therefore a differential realization of its umbral counterpart $\hat\gamma$.

3.2 Chebyshev, Lacunary Legendre and Legendre-type Polynomials

In this section, we go back to the integral representation method [3] which allows the framing of different special polynomials and functions, as well, in terms of multivariable Hermite polynomials [18] and of other polynomials belonging to standard and generalized forms of Legendre and Legendre-like type. We deal in particular with Chebyshev, Legendre, Jacobi, etc. [72].

An example of interplay between two-variable **Chebyshev polynomials of the second kind** *(CP)* $U_n(x,y)$ [3] and Hermite polynomials is provided by the following proposition.

Proposition 23. *Let*

$$U_n(x, y) = \frac{1}{n!} \int_0^\infty e^{-s} H_n(-s\,x, -sy)ds, \quad \forall x, y \in \mathbb{R}, \forall n \in \mathbb{N},$$

(3.2.1)

be the integral representation [3] of the Chebyshev polynomials of the second kind $U_n(x, y)$, then, by applying Eqs. (1.2.5)–(1.2.6)–(1.0.1), we can recast the explicit definition of $U_n(x, y)$ as

$$U_n(x, y) = (-1)^n \sum_{r=0}^{\lfloor \frac{n}{2} \rfloor} \frac{(n - r)!\, x^{n-2r}(-y)^r}{(n - 2r)!\, r!}.$$

(3.2.2)

Corollary 16. *Multiplying both sides of Eq. (3.2.1) by t^n and then by summing up over the index n, we obtain the well-known generating function of the second kind of Chebyshev polynomials, namely*

$$\sum_{n=0}^\infty t^n U_n(x, y) = \sum_{n=0}^\infty \frac{t^n}{n!} \int_0^\infty e^{-s} H_n(-s\,x, -s\,y)ds$$

$$= \int_0^\infty e^{-s\,(1+xt+yt^2)}ds = \frac{1}{1 + x\,t + y\,t^2}, \quad \mathrm{Re}(1 + x\,t + y\,t^2) > 0.$$

(3.2.3)

Using furthermore the generating function (2.4.15) [18] and employing the same procedure as before we also easily find that

$$\sum_{n=0}^\infty t^n \frac{(n + l)!}{n!} U_{n+l}(x, y) = l!\, \frac{U_l\left(\dfrac{x + 2\,y\,t}{1 + x\,t + y\,t^2}, \dfrac{y}{1 + x\,t + y\,t^2}\right)}{(1 + x\,t + y\,t^2)}$$

(3.2.4)

and indeed we get from Eq. (3.2.1)

$$\sum_{n=0}^\infty t^n \frac{(n + l)!}{n!} U_{n+l}(x, y)$$

$$= \int_0^\infty e^{-s\,(1+xt+yt^2)} H_l(-(x + 2y\,t)s, -sy)\,ds, \quad (3.2.5)$$

which, after redefining the integration variable as $s\,(1 + x\,t + y\,t^2) = \sigma$ and using again (3.2.1), yields the result reported in Eq. (3.2.3).

It is evident that the procedure we have outlined can easily be extended to generalized forms of Chebyshev polynomials $U_n^{(m)}(x,y)$ indeed, through the use of higher-order incomplete, some time also called here (even though improperly) *lacunary Hermite polynomials* (see Section 7.4), we can state

Corollary 17. *We introduce the lacunary Legendre polynomials* [110, 30] $\forall m \in \mathbb{N}, \ x, y, t \in \mathbb{R} : \ m \geq 2, \ Re(1 + xt + yt^m) > 0,$ *as follows:*

$$U_n^{(m)}(x,y) = \frac{1}{n!} \int_0^\infty e^{-s} H_n^{(m)}(-s\,x, -s\,y)ds$$

$$= (-1)^n \sum_{r=0}^{\lfloor \frac{n}{m} \rfloor} \frac{(-1)^{(m-1)r}(n - (m-1)\,r)!\,x^{n-mr}\,y^r}{(n - m\,r)!\,r!},$$

$$\sum_{n=0}^\infty t^n U_n^{(m)}(x,y) = \frac{1}{1 + x\,t + y\,t^m}.$$

$$(3.2.6)$$

For this more general case too, the use of the properties of *lacunary HP* can be usefully exploited to explore those of *lacunary CP*. With this aim, we note, e.g., the generating function in Proposition 24.

Proposition 24. *Let,* $\forall m, l \in \mathbb{N}, t \in \mathbb{R},$

$$\sum_{n=0}^\infty \frac{t^n}{n!} H_{n+l}^{(m)}(x,y) = H_l^{(m,m-1,\dots,1)}\left(\left\{\frac{p_m^{(n)}(x,y;t)}{n!}\right\}_{n=1,\dots,m}\right)$$

$$e^{p_m(x,y;\,t)},$$

$$H_n^{(p,p-1,\dots,1)}(x_1,\dots,x_p) = n! \sum_{r=0}^{\lfloor \frac{n}{p} \rfloor} \frac{H_{n-pr}^{(p-1,p-2,\dots,1)}(x_1,\dots,x_{p-1})\,x_p^r}{(n-pr)!\,r!},$$

$$p_m(x,y;t) = x\,t + y\,t^m, \quad p_m^{(n)}(x,y;t) = \partial_t^n p_m(x,y;t), \quad n \leq m$$

$$(3.2.7)$$

the Rainville generating function [72]*, where* $H_n^{(p,p-1,\dots,1)}(x_1,\dots,x_p)$ *are p-variable complete (non-lacunary) Hermite polynomials with* *generating function* [18]

$$\sum_{n=0}^{\infty} \frac{t^n}{n!} H_n^{(p,p-1,\dots,1)}(x_1,\dots,x_p) = e^{\sum_{s=1}^{p} x_s t^s}. \tag{3.2.8}$$

Then,

$$\sum_{n=0}^{\infty} t^n \frac{(n+l)!}{n!} U_{n+l}^{(m)}(x,y)$$

$$= l! \frac{U_l^{(m,m-1,\dots)} \left(\frac{p_m^{(1)}(x,y;t)}{1+p_m(x,y;t)}, \frac{1}{2} \frac{p_m^{(2)}(x,y;t)}{1+p_m(x,y;t)}, \dots, \frac{1}{m!} \frac{p_m^{(m)}(x,y;t)}{1+p_m(x,y;t)} \right)}{1+p_m(x,y;t)} \tag{3.2.9}$$

where the complete p-variable Chebyshev polynomials are specified, by means of the Laplace transform

$$U_n^{(p,p-1,\dots,1)}(x_1,\dots,x_p)$$

$$= \frac{1}{n!} \int_0^{\infty} e^{-s} H_n^{(p,p-1,\dots,1)}(-x_1 s,\dots,-x_p s) ds, \tag{3.2.10}$$

straightforwardly yielding the generating function

$$\sum_{n=0}^{\infty} t^n U_n^{(p,p-1,\dots,1)}(x_1,\dots,x_p) = \frac{1}{1+\sum_{s=1}^{p} x_s t^s},$$

$$\mathrm{Re}\left(1+\sum_{s=1}^{p} x_s t^s\right) > 0. \tag{3.2.11}$$

The formalism associated with generalized Chebyshev polynomials is fairly flexible, therefore, if we are interested in the successive derivatives of a rational function, we find the following.

Corollary 18.

$$\partial_t^m \left(\frac{1}{1+p_2(x,y;t)}\right) = \partial_t^m \int_0^{\infty} e^{-s} e^{-s\,x\,t - s\,y\,t^2} ds$$

$$= \int_0^{\infty} e^{-s} H_m\left(-s p_2^{(1)}(x,y;t), -\frac{s}{2} p_2^{(2)}(x,y;t)\right) e^{-s\,x\,t - s\,y\,t^2} ds$$

$$= m! \frac{U_m \left(\frac{p_2^{(1)}(x,y;t)}{1+p_2(x,y;t)}, \frac{1}{2} \frac{p_2^{(2)}(x,y;t)}{1+p_2(x,y;t)} \right)}{1 + p_2(x,y;\,t)}, \tag{3.2.12}$$

which yields a kind of Rodrigues formula for Chebyshev-type polynomials, as follows:

$$(1 + p_2(x,y;\,t))\, \partial_t^m \left(\frac{1}{1 + p_2(x,y;t)} \right)$$

$$= m! U_m \left(\frac{p_2^{(1)}(x,y;t)}{1 + p_2(x,y;t)}, \frac{1}{2} \frac{p_2^{(2)}(x,y;t)}{(1 + p_2(x,y;t))} \right), \tag{3.2.13}$$

which should be confronted with an analogous expression valid for the higher-order Hermite polynomials

$$\partial_t^m \left(e^{p_n(x,y;t)} \right) = H_m^{(n,n-1,\dots,1)} \left(\left\{ \frac{p_n^{(s)}(x,y;t)}{s!} \right\}_{s=1,\dots,p} \right) e^{p_n(x,y;t)}. \tag{3.2.14}$$

The formalism we have just discussed shows how the properties of Hermite polynomials and of its generalized forms is a powerful tool to deal with other families of polynomials. In the forthcoming section, we apply the obtained results to Chebyshev polynomials family.

3.2.1   *Umbral Methods and Chebyshev Polynomials*

According to the binomial umbral formalism used in Section 2.3 about *HP*, the definition of the two-variable Chebyshev polynomials can be given as follows.

Proposition 25. *We recast the two-variable Chebyshev polynomials in integral umbral binomial terms as*

$$U_n(x,y) = \frac{1}{n!} \int_0^\infty e^{-s} (-x\,s + {}_{(-y\,s)}\hat{h})^n \theta_0\, ds, \tag{3.2.15}$$

or in umbral form as

$$U_n(x,y) = \frac{1}{n!} (-x\,\hat{f} + {}_{(-y\,\hat{f})}\hat{h})^n \beta_0 \theta_0, \tag{3.2.16}$$

where $\hat{f}^r \beta_0 := \beta_r = r!$.

Proof. To prove Eq. (3.2.15) it is enough to substitute HP with binomial umbral expression (2.3.2). About Eq. (3.2.16), by using Eqs. (2.3.2), Newton binomial and $\hat{f}$-operator, we get

$$U_n(x,y) = \frac{1}{n!}\sum_{r=0}^{n}\binom{n}{r}(-x\,\hat{f})^{n-r}\left[\,_{(-y\,\hat{f})}\hat{h}\right]^r \beta_0\theta_0$$

$$= \frac{1}{n!}\sum_{r=0}^{n}\binom{n}{r}(-x)^{n-r}\,\hat{f}^{n-r}(-y\,\hat{f})^{\frac{r}{2}}\left|\cos\left(r\,\frac{\pi}{2}\right)\right|\beta_0$$

$$= \frac{1}{n!}\sum_{r=0}^{n}\binom{n}{r}(-x)^{n-r}\,\hat{f}^{n-\frac{r}{2}}(-y)^{\frac{r}{2}}\left|\cos\left(r\,\frac{\pi}{2}\right)\right|\beta_0$$

$$= (-1)^n\sum_{r=0}^{\lfloor\frac{n}{2}\rfloor}\frac{(n-r)!\,x^{n-2r}(-y)^r}{(n-2\,r)!\,r!}.$$

$\square$

Corollary 19. *According to such a definition, we find*

$$\partial_x U_n(x,y) = -\frac{\hat{f}}{(n-1)!}(-x\,\hat{f}+\,_{(-y\,\hat{f})}\hat{h})^{n-1}\beta_0. \qquad (3.2.17)$$

which yields

$$\partial_x^m U_n(x,y) = (-1)^m\frac{\hat{f}^m}{(n-m)!}(-x\,\hat{f}+\,_{(-y\,\hat{f})}\hat{h})^{n-m}\beta_0, \quad m<n,$$

$$\partial_x^n U_n(x,y) = (-1)^n n!.$$

$$(3.2.18)$$

3.2.2 *Legendre and Legendre-like Polynomials*

It is almost natural to note that the polynomials defined through the integral representation, $\forall x,y \in \mathbb{R}, \forall n \in \mathbb{N}$, [3]

$$P_n(x,y) = \frac{(-1)^n}{\sqrt{\pi}}\sum_{r=0}^{\lfloor\frac{n}{2}\rfloor}\frac{\Gamma\left(n-r+\frac{1}{2}\right)x^{n-2r}(-y)^r}{(n-2\,r)!\,r!}$$

$$= \frac{1}{n!\Gamma\left(\frac{1}{2}\right)}\int_0^{\infty}e^{-s}s^{-\frac{1}{2}}H_n(-s\,x,-sy)ds \qquad (3.2.19)$$

are generated by

$$\sum_{n=0}^{\infty} t^n P_n(x,y) = \frac{1}{\sqrt{1 + x\,t + y\,t^2}} \qquad (3.2.20)$$

and, upon replacing $x \to -2\,x$, $y = 1$, are recognized as *Legendre polynomials* [172].

All the results of the previous section can be naturally transposed to this family of polynomials thus finding the following results.

Lemma 9. $\forall m \in \mathbb{N}$

$$\partial_t^m \frac{-}{\sqrt{1 + p_2(x,y;t)}} = m! \frac{P_m \left(\dfrac{p_2^{(1)}(x,y;t)}{1 + p_2(x,y;t)}, \dfrac{1}{2!} \dfrac{p_2^{(2)}(x,y;t)}{1 + p_2(x,y;t)} \right)}{\sqrt{1 + p_2(x,y;t)}}$$

$$(3.2.21)$$

or by its extension to the lacunary forms, namely

$$\partial_t^n \frac{1}{\sqrt{1 + p_m(x,y;t)}} = n! \frac{P_n \left(\left\{ \dfrac{1}{s!} \dfrac{p_m^{(s)}(x,y;t)}{1 + p_m(x,y;t)} \right\}_{s=1,\dots,n} \right)}{\sqrt{1 + p_m(x,y;t)}} \qquad (3.2.22)$$

for the generalized case.

In the forthcoming chapter, we will complete our descrption of the use of umbral operational methods to the study of Bessel functions, thus completing the description outlined in the introductory part of the book.

Chapter 4

Bessel Functions and Umbral Calculus

In this chapter, we describe a systematic reformulation of the Theory of Special Functions, in particular of Bessel functions, in terms of the umbral conception developed so far. We begin from cylindrical Bessel functions and see how their study can be afforded by the use of elementary analytical means. We complete the program and the sparse notions we have outlined in the previous chapters, by picking an elementary function as an image of Bessel, thus drawing from its properties those of Bessel and Bessel-like forms. It is shown that computational technicalities, as, e.g., those involving the repeated derivatives of Bessel functions with respect to their variable or to their index, are indeed greatly simplified. The method is eventually extended to the wealth of Bessel function forms. Its high level of flexibility and simplicity are exploited to draw old and new elements of the underlying theory, within a very general context.

4.1 Bessel Functions

We recall the *cylindrical Bessel function of zero-order* (1.1.25) defined by the power series

$$J_0(x) = \sum_{r=0}^{\infty} \frac{(-1)^r \left(\frac{x}{2}\right)^{2r}}{(r!)^2}, \quad \forall x \in \mathbb{R}, \qquad (4.1.1)$$

whose graphical representation is given in Fig. 4.1.

The series (4.1.1) is an even continuous and infinitely differentiable function and, according to the notation outlined in Chapter 1,

93

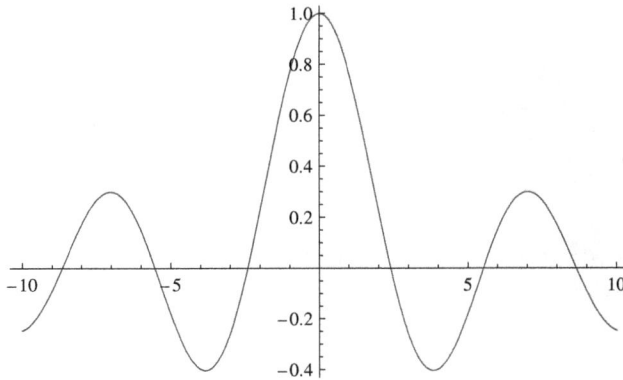

Figure 4.1:   Zeroth-order of Bessel function, $J_0(x)$.

can be written in the form of a Gaussian (1.2.17) by the use of the $\hat{c}$-operator (1.1.6)

$$J_0(x) = e^{-\hat{c}\left(\frac{x}{2}\right)^2}\varphi_0. \tag{4.1.2}$$

A non-transcendent umbral image of the Bessel function, e.g., in rational terms (1.2.25), is achieved by the use of the $\hat{b}$-operator (1.2.24)

$$J_0(x) = \frac{1}{1 + \hat{b}\left(\frac{x}{2}\right)^2}\Phi_0, \tag{4.1.3}$$

and so on.

We focus on the Gaussian umbral image for the wealth of properties that Gaussian functions possess, so allowing to derive all the properties of Bessel in a very direct way, which significantly simplifies the ordinary procedure.

The infinite integrals involving 0th order Bessel functions can be directly inferred from those of elementary Gaussian integrals. We have seen indeed (1.2.21)

$$\int_0^\infty J_0(x)dx = \left(\int_0^\infty e^{-\hat{c}\left(\frac{x}{2}\right)^2}dx\right)\varphi_0 = \sqrt{\frac{\pi}{\hat{c}}}\varphi_0 = \sqrt{\pi}\frac{1}{\Gamma\left(\frac{1}{2}\right)} = 1. \tag{4.1.4}$$

Other properties can be derived using comparable simple means.

Proposition 26. *By keeping the first derivative of both sides of Eq.* (4.1.2), *we obtain, with* $\hat{D}_x = \dfrac{d}{dx}$,

$$\hat{D}_x J_0(x) = -\left(\hat{c}\frac{x}{2}\right) e^{-\hat{c}\left(\frac{x}{2}\right)^2} \varphi_0 = -\sum_{r=0}^{\infty} \frac{(-1)^r \left(\frac{x}{2}\right)^{2r+1}}{r!(r+1)!}. \qquad (4.1.5)$$

The power series expansion in Eq. (4.1.5) *represents the* first-order cylindrical Bessel function, *namely*

$$J_1(x) = \sum_{r=0}^{\infty} \frac{(-1)^r \left(\frac{x}{2}\right)^{2r+1}}{r!(r+1)!}. \qquad (4.1.6)$$

Observation 5. We have therefore shown that

$$\hat{D}_x J_0(x) = -J_1(x). \qquad (4.1.7)$$

Corollary 20. *By keeping a further derivative of both sides of Eq.* (4.1.5), *we obtain*

$$\hat{D}_x^2 J_0(x) = -\left(\left(\hat{c}\frac{1}{2}\right) e^{-\hat{c}\left(\frac{x}{2}\right)^2} - \left(\hat{c}\frac{x}{2}\right)^2 e^{-\hat{c}\left(\frac{x}{2}\right)^2}\right) \varphi_0$$

$$= -\frac{J_1(x)}{x} + \sum_{r=0}^{\infty} \frac{(-1)^r \left(\frac{x}{2}\right)^{2r+2}}{r!(r+2)!}. \qquad (4.1.8)$$

The power series in the last equation provides the second-order Bessel equation, *namely*

$$J_2(x) = \sum_{r=0}^{\infty} \frac{(-1)^r \left(\frac{x}{2}\right)^{2r+2}}{r!(r+2)!}. \qquad (4.1.9)$$

More in general, by keeping successive derivatives (see below for a more complete treatment), we can argue that the nth order *of Bessel functions is*

$$J_n(x) = \left(\hat{c}\frac{x}{2}\right)^n e^{-\hat{c}\left(\frac{x}{2}\right)^2} \varphi_0 = \sum_{r=0}^{\infty} \frac{(-1)^r \left(\frac{x}{2}\right)^{2r+n}}{r!(r+n)!}. \qquad (4.1.10)$$

Observation 6. We note that we can obtain the same result by keeping the n-th derivative with respect to $\left(\frac{x}{2}\right)^2$ so, $\forall x \in \mathbb{R}, \forall n \in \mathbb{N}$,

$$\left(\frac{d}{d\left[\left(\frac{x}{2}\right)^2\right]}\right)^n J_0(x) = (-1)^n \hat{c}^n e^{-\hat{c}\left(\frac{x}{2}\right)^2} \varphi_0 = (-1)^n \sum_{r=0}^{\infty} \frac{(-1)^r \left(\frac{x}{2}\right)^{2r}}{r!(n+r)!}$$

(4.1.11)

or

$$\left(\frac{1}{x}\frac{d}{dx}\right)^n J_0(x) = \sum_{r=0}^{\infty} \frac{(-1)^r \left(\frac{x}{2}\right)^{2r}}{r!(n+r)!}$$

(4.1.12)

and finally

$$J_n(x) = \left(\frac{x}{2}\right)^n \left(\frac{1}{x}\frac{d}{dx}\right)^n J_0(x) = \sum_{r=0}^{\infty} \frac{(-1)^r \left(\frac{x}{2}\right)^{2r+n}}{r!(n+r)!}$$

$$= \left(\hat{c}\frac{x}{2}\right)^n e^{-\hat{c}\left(\frac{x}{2}\right)^2} \varphi_0.$$

(4.1.13)

The above relation can be further exploited to get a general recurrence relation involving contiguous index Bessel functions and their derivatives.

Properties 12. From Eq. (4.1.10), we find

$$\text{(i)} \; \hat{D}_x J_n(x) = \left(\frac{n}{x}\left(\hat{c}\frac{x}{2}\right)^n - \left(\hat{c}\frac{x}{2}\right)^{n+1}\right) e^{-\hat{c}\left(\frac{x}{2}\right)^2} \varphi_0$$

$$= \frac{n}{x} J_n(x) - J_{n+1}(x).$$

(4.1.14)

The use of the series expansion yields

$$\text{(ii)} \; n J_n(x) = n \sum_{r=0}^{\infty} \frac{(-1)^r \left(\frac{x}{2}\right)^{2r+n}}{r!(r+n)!}$$

$$= \sum_{r=0}^{\infty} \frac{(-1)^r (n+r) \left(\frac{x}{2}\right)^{2r+n}}{r!(r+n)!} - \sum_{r=0}^{\infty} \frac{(-1)^r r \left(\frac{x}{2}\right)^{2r+n}}{r!(r+n)!}$$

$$= \frac{x}{2} \left(J_{n-1}(x) + J_{n+1}(x)\right)$$

(4.1.15)

which, for convenience, is rewritten in the form of the recurrence

$$\frac{n}{x} J_n(x) = \frac{1}{2} \left(J_{n-1}(x) + J_{n+1}(x) \right), \tag{4.1.16}$$

exploited, along with Eq. (4.1.14), to get the identity

$$\hat{D}_x J_n(x) = \frac{1}{2} \left(J_{n-1}(x) - J_{n+1}(x) \right). \tag{4.1.17}$$

It is also straightforwardly proved that

$$\text{(iii)} \ \hat{D}_x \left(x^n J_n(x) \right) = \hat{D}_x \left(\left(\frac{\hat{c}}{2} \right)^n x^{2n} e^{-\hat{c} \left(\frac{x}{2} \right)^2} \varphi_0 \right) = x^n J_{n-1}. \tag{4.1.18}$$

Corollary 21. *The differential equation satisfied by the Bessel functions can be obtained from the previous recurrences* (4.1.16)– (4.1.17) *by noting that, once combined, they yield*

$$\begin{aligned} J_{n-1}(x) &= \frac{n}{x} J_n(x) + \hat{D}_x J_n(x), \\ J_{n+1}(x) &= \frac{n}{x} J_n(x) - \hat{D}_x J_n(x). \end{aligned} \tag{4.1.19}$$

Proposition 27. *We introduce the index shifting operators*

$$\hat{E}_\pm = \frac{\hat{N}}{x} \mp \hat{D}_x, \tag{4.1.20}$$

with $\hat{N}$ being an "index" operator defined as

$$\hat{N} J_n(x) = n J_n(x). \tag{4.1.21}$$

Then we can express the recurrence relations (4.1.19) *in the form*

$$\hat{E}_\pm J_n(x) = J_{n \pm 1}(x). \tag{4.1.22}$$

It is also easily understood that the action of the product of the shift operators yield

$$\hat{E}_+ \hat{E}_- \left(J_n(x) \right) = J_n(x), \tag{4.1.23}$$

which in differential forms reads

$$\hat{E}_+\hat{E}_-\left(J_n(x)\right) = \left(\frac{\hat{N}}{x} - \hat{D}_x\right)J_{n-1}(x) = \left(\frac{n-1}{x} - \hat{D}_x\right)J_{n-1}(x)$$

$$= \left(\frac{n-1}{x} - \hat{D}_x\right)\left(\frac{n}{x} + \hat{D}_x\right)J_n(x) \qquad (4.1.24)$$

and, after a few manipulations, we end up with

$$\left\{x^2\hat{D}_x^2 + x\hat{D}_x + \left(x^2 - n^2\right)\right\}J_n(x) = 0 \qquad (4.1.25)$$

or, in a more compact form,

$$\left(x\hat{D}_x\right)^2 J_n(x) = (n^2 - x^2)J_n(x). \qquad (4.1.26)$$

The products of two Bessel functions (which will be detailed successively) provide new families of functions with interesting properties. Identities like the recurrence relations can be extended to, e.g., the *square of Bessel functions*.

Proposition 28.

(a) $\left(\frac{1}{2}\left(\hat{D}_x^2 + \frac{1}{x}\hat{D}_x\right) + 1\right)J_n^2(x) = \frac{1}{2}\left(J_{n-1}^2(x) + J_{n+1}^2(x)\right),$

(b) $\frac{n}{x}\hat{D}_x J_n^2(x) = \frac{1}{2}\left(J_{n-1}^2(x) - J_{n+1}^2(x)\right).$

$$(4.1.27)$$

Proof. Equation (b) follows by noting that

$$\frac{n}{x}\hat{D}_x J_n^2(x) = \left(\frac{2n}{x}J_n(x)\right)\left(\hat{D}J_n(x)\right), \qquad (4.1.28)$$

which, on account of the recurrences (4.1.16)–(4.1.17), yields

$$\frac{n}{x}\hat{D}_x J_n^2(x) = \frac{1}{2}\left(J_{n-1}(x) + J_{n+1}(x)\right)\left(J_{n-1}(x) - J_{n+1}(x)\right). \quad (4.1.29)$$

The derivation of the $(4.1.27)-a))$ is elaborate more as follows:

(i) Note that from Eqs. (4.1.16) and (4.1.17) it follows that

$$\left(\frac{n}{x}\right)^2 J_n^2(x) = \frac{1}{4}\left(J_{n+1}^2(x) + J_{n-1}^2(x) + 2J_{n+1}(x)J_{n-1}(x)\right),$$

$$\left(\hat{D}_x J_n(x)\right)^2 = \frac{1}{4}\left(J_{n+1}^2(x) + J_{n-1}^2(x) - 2J_{n+1}(x)J_{n-1}(x)\right).$$

$$(4.1.30)$$

(ii) Apply standard operatorial rules to state the identity

$$\left(\hat{D}_x J_n(x)\right)^2 = \frac{1}{2}\hat{D}_x^2 J_n^2(x) - J_n(x)\hat{D}_x^2 J_n(x)$$

$$= \frac{1}{2}\hat{D}_x^2 J_n^2(x) - \frac{J_n(x)}{x^2}(x^2 \hat{D}_x^2 J_n(x)). \qquad (4.1.31)$$

(iii) Use the Bessel differential equation to get

$$\frac{1}{2}\hat{D}_x^2 J_n^2(x) + \frac{1}{2x}\hat{D}_x J_n^2(x) + \left(1 - \frac{n^2}{x^2}\right) J_n^2(x)$$

$$= \frac{1}{4}\left[J_{n+1}^2(x) + J_{n-1}^2(x) - 2J_{n+1}(x)J_{n-1}(x)\right], \qquad (4.1.32)$$

which combined with the first in (4.1.30) yields the first of the recurrences (4.1.27). □

Corollary 22. *The use of the shift operator formalism can also be exploited to derive the differential equation satisfied by $J_n^2(x)$.*

It is indeed fairly natural, in analogy with (4.1.22), to set

$$_2\hat{E}_- J_n^2(x) = J_{n-1}^2(x), \quad _2\hat{E}_+ J_n^2(x) = J_{n+1}^2(x), \qquad (4.1.33)$$

where the shift operators have been defined as

$$_2\hat{E}_- = \left[\frac{1}{2}\left(\hat{D}_x^2 + \frac{1}{x}\hat{D}_x\right) + 1\right] + \frac{\hat{N}}{x}\hat{D}_x,$$

$$\qquad (4.1.34)$$

$$_2\hat{E}_+ = \left[\frac{1}{2}\left(\hat{D}_x^2 + \frac{1}{x}\hat{D}_x\right) + 1\right] - \frac{\hat{N}}{x}\hat{D}_x.$$

The differential equation satisfied by the function $y(x) = J_n^2(x)$ is, thereby, provided by

$$_2\hat{E}_- \left(_2\hat{E}_+ y \right) = y,$$ (4.1.35)

namely

$$\left(\hat{\Theta} + \frac{n+1}{x} \hat{D}_x \right) \left(\hat{\Theta} - \frac{n}{x} \hat{D}_x \right) y = y,$$

$$\hat{\Theta} = \frac{1}{2} \left(\hat{D}_x^2 + \frac{1}{x} \hat{D}_x \right) + 1,$$ (4.1.36)

which is an jourth-order differential equation which is worth expanding as the form[a]

$$\left(\hat{\Theta}^2 + \frac{1}{x} \hat{D}_x \hat{\Theta} + \frac{n(n+1)}{x^2} \left(\frac{1}{x} - \hat{D}_x \right) \hat{D}_x - n \left[\hat{\Theta}, \frac{1}{x} \hat{D}_x \right] \right) y = y.$$ (4.1.37)

The explicit evaluation of the commutator yields

$$\left[\hat{\Theta}, \frac{1}{x} \hat{D}_x \right] = \hat{\Theta} \frac{1}{x} \hat{D}_x - \frac{1}{x} \hat{D}_x \hat{\Theta} = \frac{1}{x^2} \left(\frac{1}{x} - \hat{D}_x \right) \hat{D}_x$$ (4.1.38)

and Eq. (4.1.37) becomes

$$\left(\hat{\Theta}^2 + \frac{1}{x} \hat{D}_x \hat{\Theta} + n^2 \left[\hat{\Theta}, \frac{1}{x} \hat{D}_x \right] \right) y = y,$$ (4.1.39)

thus eventually getting

$$\left(x^3 \hat{\Theta}^2 + x^2 \hat{D}_x \hat{\Theta} + n^2 (1 - x \hat{D}_x) \hat{D}_x - x^3 \right) y = 0.$$ (4.1.40)

The form we have just derived is different from the analogous expression given in Abramowitz and Stegun [1], where the relevant differential equation is fourth order in terms of the operator $\hat{\vartheta} = x \hat{D}_x$. It is possible to compare Eqs. (4.1.39)–(4.1.40) with the form given in Abramowitz and Stegun and extend the previous results to products of Bessel functions with different indices.

---

[a]The bracket $[\hat{A}, \hat{B}] = \hat{A}\hat{B} - \hat{B}\hat{A}$ denotes the commutation parenthesis and is different from zero whenever $\hat{A}, \hat{B}$ are not commuting quantities.

Further examples of application of the just outlined method are as follows.

Example 36. Given the function $\phi_n(x) = x^p J_n(\lambda x^q)$ with $\forall n \in \mathbb{N}$ and λ, p, q independent of the index n, we can find the relevant recurrences and differential equation.

We note that

$$\frac{2n}{\lambda x^q} J_n(\lambda x^q) = J_{n-1}(\lambda x^q) + J_{n+1}(\lambda x^q). \qquad (4.1.41)$$

We multiply both sides by x^p and deduce that

$$\frac{2n}{\lambda x^q} \phi_n(x) = \phi_{n-1}(x) + \phi_{n+1}(x). \qquad (4.1.42)$$

By noting that

$$J_n(\lambda x^q) = x^{-p}\phi_n(x) \qquad (4.1.43)$$

and by using the fact

$$2\hat{D}_Z J_n(Z) = J_{n-1}(Z) - J_{n+1}(Z), \quad Z = \lambda x^q \qquad (4.1.44)$$

and, after establishing the identity

$$\hat{D}_Z = \frac{1}{q\lambda x^{q-1}} \hat{D}_x, \qquad (4.1.45)$$

we end up with

$$(p + nq)\phi_{n-1}(x) + (p - nq)\phi_{n+1}(x) = \frac{2n}{\lambda} x^{1-q} \phi'_n(x). \qquad (4.1.46)$$

The relevant differential equations are obtained either by using an appropriate generalization of the shift operator method, or by noting that

$$[Z^2 \hat{D}_Z^2 + Z\hat{D}_Z + (Z^2 - n^2)](x^{-p}\phi_n(x)) = 0. \qquad (4.1.47)$$

Other properties which can be easily checked using the present formalism are as follows.

Properties 13. We have:

(a) Argument Reflection Condition

$$J_n(-x) = \left(\hat{c}\frac{(-x)}{2}\right)^n e^{-\hat{c}\left(\frac{x}{2}\right)^2}\varphi_0 = (-1)^n J_n(x). \qquad (4.1.48)$$

(b) Index Reflection Condition

$$J_{-n}(x) = \left(\hat{c}\frac{x}{2}\right)^{-n} e^{-\hat{c}\left(\frac{x}{2}\right)^2}\varphi_0 = \sum_{r=n}^{\infty} \frac{(-1)^r \left(\frac{x}{2}\right)^{2r-n}}{r!(r-n)!}$$

$$= \sum_{s=0}^{\infty} \frac{(-1)^{s+n}\left(\frac{x}{2}\right)^{2s+n}}{s!(s+n)!} = (-1)^n J_n(x). \qquad (4.1.49)$$

Example 37. The use of the umbral formalism makes the derivation of the sum rule straightforward

$$\sum_{n=0}^{\infty} \frac{t^n}{n!} J_n(x) = \sum_{n=0}^{\infty} \frac{1}{n!}\left(t\,\hat{c}\frac{x}{2}\right)^n e^{-\hat{c}\left(\frac{x}{2}\right)^2}\varphi_0 = e^{-\frac{\hat{c}}{4}\left(x^2-2xt\right)}\varphi_0$$

$$= J_0\left(\sqrt{x^2 - 2xt}\right). \qquad (4.1.50)$$

The extension of the previous formula to a slightly more complicated case involving $J_{n+l}(x)$ is obtained along the same straightforward line

$$\sum_{n=0}^{\infty} \frac{t^n}{n!} J_{n+l}(x) = \left(\hat{c}\frac{x}{2}\right)^l \sum_{n=0}^{\infty} \frac{t^n}{n!} \left(\hat{c}\frac{x}{2}\right)^n e^{-\hat{c}\left(\frac{x}{2}\right)^2}\varphi_0$$

$$= \left(\hat{c}\frac{x}{2}\right)^l e^{-\frac{\hat{c}}{4}\left(x^2-2xt\right)}\varphi_0$$

$$= \left(\frac{x}{x-2t}\right)^{\frac{l}{2}} J_l\left(\sqrt{x^2 - 2xt}\right). \qquad (4.1.51)$$

Further comments will be given in Chapter 7.

In this section, we have derived the main properties of Bessel functions in a quite straightforward way and most of the simplicity of the results are associated with the formalism we used.

4.2 Bessel Functions of the Second Kind

In the previous section, we have considered integer order Bessel functions, but there is no argument against their extension to any *Real* order, namely

Definition 15. We introduce the Gaussian umbral form of *real order Bessel function*

$$J_\nu(x) = \left(\hat{c}\frac{x}{2}\right)^\nu e^{-\hat{c}\left(\frac{x}{2}\right)^2}\varphi_0 = \sum_{r=0}^{\infty} \frac{(-1)^r \left(\frac{x}{2}\right)^{2r+\nu}}{r!\,\Gamma(\nu+r+1)}, \quad \forall \nu \in \mathbb{R}. \quad (4.2.1)$$

It is easily checked that the function (4.2.1) satisfies the same recurrences of the integer order case and, therefore, the same differential equation.

Corollary 23. *From Eq. (4.2.1) we also find*

$$J_\nu(-x) = \left(-\hat{c}\frac{x}{2}\right)^\nu e^{-\hat{c}\left(\frac{x}{2}\right)^2}\varphi_0 = e^{i\pi\nu}J_\nu(x). \quad (4.2.2)$$

It is now worth noting that as (4.1.25) is a second-order differential equation, it admits two independent solutions. In the case of integer index $J_n(x), J_{-n}(x)$ cannot be considered independent, but, since the non-integer case $J_\nu(x)$ is not expressible in terms of $J_{-\nu}(x)$ and vice versa, we can consider them as independent solutions of the Bessel equation.

Any linear combination of the two solutions represents a solution of the Bessel equation, therefore, we set [3] the following definition and show the subsequent theorem.

Definition 16. The identity

$$Y_\nu(x) = \frac{J_\nu(x)\cos(\nu\pi) - J_{-\nu}(x)}{\sin(\nu\pi)} \quad (4.2.3)$$

is called the *Neumann-Bessel* (NB) function.

Theorem 3. *Let $\forall x \in \mathbb{R}$, $\forall \nu \in \mathbb{R}$, $Y_\nu(x)$ be the NB function, then $\forall n \in \mathbb{Z}$,*

$$\lim_{\nu \to n} Y_\nu(x) = Y_n(x), \tag{4.2.4}$$

where $Y_n(x)$ is the integer index counterpart of (4.2.3).

Proof. According to our definition (4.2.1), we obtain

$$Y_\nu(x) = \frac{\left(\frac{\hat{c}x}{2}\right)^\nu \cos(\nu\pi) - \left(\frac{\hat{c}x}{2}\right)^{-\nu}}{\sin(\nu\pi)} e^{-\hat{c}\left(\frac{x}{2}\right)^2} \varphi_0 \tag{4.2.5}$$

so, for (4.1.10) and (4.1.49),

$$\lim_{\nu \to n} Y_\nu(x) = \lim_{\nu \to n} \frac{\left(\frac{\hat{c}x}{2}\right)^\nu \cos(\nu\pi) - \left(\frac{\hat{c}x}{2}\right)^{-\nu}}{\sin(\nu\pi)} e^{-\hat{c}\left(\frac{x}{2}\right)^2} \varphi_0$$

$$= \frac{\pm J_n(x) - J_{-n}(x)}{0} = \frac{0}{0}. \tag{4.2.6}$$

Applying the de L'Hopital rule, the derivation with respect to ν yields

$$\lim_{\nu \to n} Y_\nu(x)$$

$$= \lim_{\nu \to n} \frac{\left(\hat{c}\frac{x}{2}\right)^\nu \ln\left(\hat{c}\frac{x}{2}\right)\cos(\nu\pi) - \pi\left(\hat{c}\frac{x}{2}\right)^\nu \sin(\nu\pi) + \left(\hat{c}\frac{x}{2}\right)^{-\nu}\ln\left(\hat{c}\frac{x}{2}\right)}{\pi\cos(\nu\pi)}$$

$$\cdot e^{-\hat{c}\left(\frac{x}{2}\right)^2}\varphi_0 = \frac{\ln\left(\frac{x}{2}\right)\left((-1)^n\left(\hat{c}\frac{x}{2}\right)^n + \left(\hat{c}\frac{x}{2}\right)^{-n}\right)}{(-1)^n\pi} e^{-\hat{c}\left(\frac{x}{2}\right)^2}\varphi_0$$

$$+ \frac{1}{(-1)^n\pi}\lim_{\nu \to n}\ln(\hat{c})\left((-1)^\nu\left(\hat{c}\frac{x}{2}\right)^\nu + \left(\hat{c}\frac{x}{2}\right)^{-\nu}\right)e^{-\hat{c}\left(\frac{x}{2}\right)^2}\varphi_0$$

$$= \frac{1}{\pi}\ln\left(\frac{x}{2}\right)J_n(x) + \frac{1}{(-1)^n\pi}\ln\left(\frac{x}{2}\right)J_{-n}(x)$$

$$+ \frac{1}{(-1)^n\pi}\lim_{\nu \to n}\ln(\hat{c})\left((-1)^\nu\left(\hat{c}\frac{x}{2}\right)^\nu + \left(\hat{c}\frac{x}{2}\right)^{-\nu}\right)e^{-\hat{c}\left(\frac{x}{2}\right)^2}\varphi_0$$

$$= \frac{2}{\pi} \ln \left(\frac{x}{2} \right) J_n(x) + \lim_{\nu \to n} \frac{1}{\pi} \sum_{r=0}^{\infty} \frac{(-1)^r \left(\frac{x}{2} \right)^{2r+\nu} \hat{c}^{r+\nu} \ln(\hat{c})}{r!} \varphi_0$$

$$+ \frac{1}{(-1)^n \pi} \lim_{\nu \to n} \sum_{r=0}^{\infty} \frac{(-1)^r \left(\frac{x}{2} \right)^{2r-\nu} \hat{c}^{r-\nu} \ln(\hat{c})}{r!} \varphi_0. \qquad (4.2.7)$$

Now, to evaluate the logarithm of the operator $\hat{c}$ on φ_0, we use (1.1.6) and set

$$\ln(\hat{c})\varphi_0 = \lim_{\alpha \to 0} \left(\frac{\hat{c}^\alpha - 1}{\alpha} \right) \varphi_0 = \lim_{\alpha \to 0} \frac{\varphi_\alpha - \varphi_0}{\alpha}$$

$$= \lim_{\alpha \to 0} \left[\frac{1}{\alpha} \left(\frac{1}{\Gamma(\alpha + 1)} - 1 \right) \right] = - \lim_{\alpha \to 0} \frac{\Gamma'(\alpha + 1)}{(\Gamma(\alpha + 1))^2} =_{de\ L'H.}$$

$$= - \lim_{\alpha \to 0} \frac{\Psi(\alpha + 1)}{\Gamma(\alpha + 1)}, \qquad (4.2.8)$$

where

$$\Psi(\beta) = \frac{\Gamma'(\beta)}{\Gamma(\beta)}, \quad \beta \in \mathbb{R}, \qquad (4.2.9)$$

is the *Digamma* function (see Chapter 7). Then,

$$\lim_{\nu \to n} Y_\nu(x) = \frac{2}{\pi} \ln \left(\frac{x}{2} \right) J_n(x) + \frac{1}{\pi} \sum_{r=0}^{\infty} \frac{(-1)^r \left(\frac{x}{2} \right)^{2r+n}}{r!}$$

$$\cdot \lim_{\alpha \to 0} \left(\frac{\hat{c}^{\alpha+r+n} - \hat{c}^{r+n}}{\alpha} \right) \varphi_0 + \frac{1}{(-1)^n \pi} \sum_{r=0}^{\infty} \frac{(-1)^r \left(\frac{x}{2} \right)^{2r-n}}{r!}$$

$$\cdot \lim_{\nu \to n} \lim_{\alpha \to 0} \left(\frac{\hat{c}^{\alpha+r-\nu} - \hat{c}^{r-\nu}}{\alpha} \right) \varphi_0, \qquad (4.2.10)$$

which becomes

$$\lim_{\nu \to n} Y_\nu(x)$$

$$= \frac{2}{\pi} \ln \left(\frac{x}{2} \right) J_n(x) - \frac{1}{\pi} \sum_{r=0}^{\infty} \frac{(-1)^r \left(\frac{x}{2} \right)^{2r+n}}{r!} \lim_{\alpha \to 0} \frac{\Psi(\alpha + r + n + 1)}{\Gamma(\alpha + r + n + 1)}$$

$$+ \frac{1}{(-1)^n \pi} \left[- \sum_{r=0}^{n-1} \frac{(-1)^r \left(\frac{x}{2}\right)^{2r-n}}{r!} \lim_{\nu \to n} \frac{\Psi(r - \nu + 1)}{\Gamma(r - \nu + 1)} \right.$$

$$\left. + \sum_{r=n}^{\infty} \frac{(-1)^r \left(\frac{x}{2}\right)^{2r-n}}{r!} \lim_{\nu \to n} \lim_{\alpha \to 0} \left(\frac{\hat{c}^{\alpha+r-\nu} - \hat{c}^{r-\nu}}{\alpha} \right) \right] \varphi_0$$

$$= \frac{2}{\pi} \ln \left(\frac{x}{2}\right) J_n(x) - \frac{1}{\pi} \sum_{r=0}^{\infty} \frac{(-1)^r \left(\frac{x}{2}\right)^{2r+n}}{r!} \frac{\Psi(r + n + 1)}{\Gamma(r + n + 1)}$$

$$+ \frac{1}{(-1)^n \pi} \left[- \sum_{r=0}^{n-1} \frac{(-1)^r \left(\frac{x}{2}\right)^{2r-n}}{r!} \lim_{\nu \to n} \frac{\Psi(r - \nu + 1)}{\Gamma(r - \nu + 1)} \right.$$

$$\left. + \sum_{s=0}^{\infty} \frac{(-1)^s \left(\frac{x}{2}\right)^{2s+n}}{(n+s)!} \lim_{\alpha \to 0} \left(\frac{\hat{c}^{\alpha+s} - \hat{c}^s}{\alpha} \right) \varphi_0 \right], \tag{4.2.11}$$

and finally

$$\lim_{\nu \to n} Y_\nu(x)$$

$$= \frac{2}{\pi} \ln \left(\frac{x}{2}\right) J_n(x) - \frac{1}{\pi} \sum_{r=0}^{\infty} \frac{(-1)^r \left(\frac{x}{2}\right)^{2r+n}}{r!} \frac{\Psi(r + n + 1)}{\Gamma(r + n + 1)}$$

$$+ \frac{1}{(-1)^n \pi} \left[- \sum_{r=0}^{n-1} \frac{(-1)^r \left(\frac{x}{2}\right)^{2r-n}}{r!} \lim_{\nu \to n} \frac{\Psi(r - \nu + 1)}{\Gamma(r - \nu + 1)} \right.$$

$$\left. - \sum_{s=0}^{\infty} \frac{(-1)^{n+s} \left(\frac{x}{2}\right)^{2s+n}}{(n+s)!} \frac{\Psi(s + 1)}{\Gamma(s + 1)} \right] \tag{4.2.12}$$

but, using the identities

$$\Gamma(x)\Gamma(1 - x) = \frac{\pi}{\sin(x\pi)}, \quad \forall x \notin \mathbb{Z}, \tag{4.2.13}$$

$$\Psi(1 - x) - \Psi(x) = \pi \cot(x\pi), \quad \forall x \notin \mathbb{Z}, \tag{4.2.14}$$

we can write

$$\lim_{\nu \to n} \frac{\Psi(1 - (\nu - r))}{\Gamma(1 - (\nu - r))} = \begin{cases} (-1)^{n-r}(n - r - 1)! & 0 \le r < n, \\ -\infty & r \ge n. \end{cases} \tag{4.2.15}$$

We eventually find

$$\lim_{\nu \to n} Y_\nu(x) = \frac{2}{\pi} \ln\left(\frac{x}{2}\right) J_n(x) - \frac{1}{\pi} \sum_{r=0}^{\infty} \frac{(-1)^r \left(\frac{x}{2}\right)^{2r+n}}{r!} \frac{\Psi(r+n+1)}{\Gamma(r+n+1)}$$

$$- \frac{1}{\pi} \left[\sum_{r=0}^{n-1} \frac{(n-r-1)!}{r!} \left(\frac{x}{2}\right)^{2r-n} \right.$$

$$\left. + \sum_{s=0}^{\infty} \frac{(-1)^s}{(n+s)!} \frac{\Psi(s+1)}{s!} \left(\frac{x}{2}\right)^{2s+n} \right]$$

$$= Y_n(x). \tag{4.2.16}$$

□

Corollary 24. *Now, we recast in operators terms and obtain*

$$\Delta_n(x) = \left(\hat{\chi}\hat{c}\frac{x}{2}\right)^n e^{-\hat{\chi}\hat{c}\left(\frac{x}{2}\right)^2} \varphi_0 \vartheta_0 = \sum_{k=0}^{\infty} \frac{(-1)^k \Psi(k+n+1)}{k!\Gamma(k+n+1)} \left(\frac{x}{2}\right)^{2k+n},$$

$$\hat{\chi}^\alpha \vartheta_0 = \frac{\Gamma'(\alpha+1)}{\Gamma(\alpha+1)},$$

$$\delta_n(x) = \left(\hat{c}\frac{x}{2}\right)^n e^{-\hat{\chi}\hat{c}\left(\frac{x}{2}\right)^2} \varphi_0 \vartheta_0 = \sum_{k=0}^{\infty} \frac{(-1)^k \Psi(k+1)}{k!\Gamma(k+n+1)} \left(\frac{x}{2}\right)^{2k+n}$$

$$\tag{4.2.17}$$

and these positions yield

$$Y_n(x) = \frac{2}{\pi} \ln\left(\frac{x}{2}\right) J_n(x) - \frac{1}{\pi} \sum_{k=0}^{n-1} \frac{(n-k-1)!}{k!} \left(\frac{x}{2}\right)^{2k-n}$$

$$- \frac{1}{\pi} [\Delta_n(x) + \delta_n(x)] \tag{4.2.18}$$

which is the nth order of NB function. In particular, the 0th order is

$$Y_0(x) = \frac{2}{\pi} \ln\left(\frac{x}{2}\right) J_0(x) - \frac{2}{\pi} \Delta_0(x). \tag{4.2.19}$$

The function $Y_0(x)$ has a logarithmic singularity at the origin and is the second solution of the Bessel equation with $n = 0$.

Observation 7. The functions $\Delta_n(x), \delta_n(x)$ are not usually considered as independent functions. We call them the first and second renormalized Bessel functions, respectively, for the reasons we will clarify in the following and we will discuss their properties later in this book.

Corollary 25. *The theorem provides the derivative with respect to* ν *of the Bessel function*

$$\hat{D}_\nu J_\nu(x) = \left(\hat{c}\frac{x}{2}\right)^\nu e^{-\hat{c}\left(\frac{x}{2}\right)^2} \left[\ln\left(\frac{x}{2}\right) + \ln(\hat{c})\right] \varphi_0$$

$$= \ln\left(\frac{x}{2}\right) J_\nu(x) - \sum_{r=0}^{\infty} \frac{(-1)^r \Psi(r+\nu+1)}{r!\,\Gamma(r+\nu+1)} \left(\frac{x}{2}\right)^{2r+\nu} .$$

$$(4.2.20)$$

The forthcoming section is addressed to the modified form of Bessel functions.

4.3 Modified Bessel Functions of the First Kind

The *modified Bessel functions of the first kind* are a byproduct of the ordinary cylinder Bessel functions and can be defined as follows.

Definition 17. We introduce the umbral form of modified Bessel functions of the first kind

$$I_\nu(x) = \left(\hat{c}\frac{x}{2}\right)^\nu e^{\hat{c}\left(\frac{x}{2}\right)^2} \varphi_0 = \sum_{r=0}^{\infty} \frac{\left(\frac{x}{2}\right)^{2r+\nu}}{r!\,\Gamma(\nu+r+1)}. \qquad (4.3.1)$$

The link of the modified Bessel functions of the first kind with the $J_\nu(x)$ counterpart is provided by the following property.

Properties 14.

$$I_\nu(ix) = i^\nu \left(\hat{c}\frac{x}{2}\right)^\nu e^{-\hat{c}\left(\frac{x}{2}\right)^2} \varphi_0 = e^{i\nu\frac{\pi}{2}} J_\nu(x). \qquad (4.3.2)$$

Proposition 29. *The differential equation they satisfy can be inferred from (4.1.25) by replacing x with ix, therefore, we find*

$$\left\{ x^2 \hat{D}_x^2 + x \hat{D}_x - \left[x^2 + \nu^2 \right] \right\} I_\nu(x) = 0. \tag{4.3.3}$$

The second solution is a Bessel-like function too, usually referred as Macdonald function or Modified Bessel function of the second kind, defined as

$$K_\nu(x) = \frac{\pi}{2} \frac{I_{-\nu}(x) - I_\nu(x)}{\sin(\nu\pi)}$$

$$= -\frac{\pi}{\sin(\nu\pi)} \left[\sinh\left(\nu \ln\left(\frac{\hat{c}x}{2} \right) \right) \right] e^{\hat{c}\left(\frac{x}{2} \right)^2} \varphi_0. \tag{4.3.4}$$

The integer order counterpart can be obtained by following a procedure analogous to that leading to the integer order of Macdonald functions. We get indeed

$$K_n(x) = -\lim_{\nu \to \pi} \frac{1}{\cosh(\nu\pi)} \left[\cosh\left[\nu \ln\left(\frac{\hat{c}x}{2} \right) \right] \ln\left(\frac{\hat{c}x}{2} \right) \right] e^{\hat{c}\left(\frac{x}{2} \right)^2} \varphi_0$$

$$= (-1)^{n+1} \left[\cosh\left[n \ln\left(\frac{\hat{c}x}{2} \right) \right] \ln\left(\frac{\hat{c}x}{2} \right) \right] e^{\hat{c}\left(\frac{x}{2} \right)^2} \varphi_0. \tag{4.3.5}$$

In the case $n = 0$, we obtain

$$K_0(x) = - \left[\ln\left(\frac{x}{2} \right) + \ln(\hat{c}) \right] e^{\hat{c}\left(\frac{x}{2} \right)^2} \varphi_0$$

$$= - \ln\left(\frac{x}{2} \right) I_0(x) - \lim_{\alpha \to 0} \frac{\hat{c}^\alpha - 1}{\alpha} e^{\hat{c}\left(\frac{x}{2} \right)^2} \varphi_0$$

$$= - \ln\left(\frac{x}{2} \right) I_0(x) + \sum_{r=0}^{\infty} \frac{\left(\frac{x}{2} \right)^{2r}}{r!^2} \Psi(r + 1), \tag{4.3.6}$$

and, therefore, by introducing the $\hat{h}$-operator,

$$K_0(x) = - \left[\ln\left(\frac{x}{2} \right) + \gamma \right] I_0(x) + I_0(\hat{h}^{\frac{1}{2}} x),$$

$$I_0(\hat{h}x) = \sum_{r=0}^{\infty} \frac{\left(\frac{\hat{h}x}{2} \right)^{2r}}{(r!)^2}, \tag{4.3.7}$$

with

$$\hat{h}^0 = 0, \quad \hat{h}^m = h_m, \quad h_m = \sum_{k=1}^{m} \frac{1}{k}, \tag{4.3.8}$$

where h_m are the Harmonic Numbers.[b]

This section completes a first elementary analysis of the Bessel functions. The remaining parts of the chapter deal with further Bessel-like forms, including spherical, Strüve and multivariable cases.

4.4 Umbra and Spherical/Strüve Bessel

In the previous sections, we have provided our "non-ordinary" point of view to the theory of "ordinary" Bessel functions. This section offers the same treatment for the study of other Bessel-families, often occurring in applications.

The spherical Bessels are widely exploited in physical problems, concerning, for example, diffraction and scattering of radiation [106] or scattering problems in Quantum Mechanics [23]. Although their properties are widely well known, operational methods, developed so far, will give an effective and concise tool for their framing within a coherent context.

The cylindrical Bessel functions are linked to their spherical counterparts by

$$j_n(x) = \sqrt{\frac{\pi}{2x}} J_{n+\frac{1}{2}}(x). \tag{4.4.1}$$

The differential equation they satisfy can be obtained by the use of a technique analogous to that exploited for the first and second kinds of cylindrical Bessels.

We first note that, in terms of shift operators $\hat{E}_{\pm}$ we find,

$$\hat{E}_+ \hat{E}_- \left(\sqrt{\frac{2x}{\pi}} j_n(x) \right) = J_{n+\frac{1}{2}}(x) \tag{4.4.2}$$

---

[b]They will be treated in Chapter 6.

which, in differential form, reads

$$\left(\frac{n-\frac{1}{2}}{x} - \hat{D}_x\right)\left(\frac{n+\frac{1}{2}}{x} - \hat{D}_x\right)\left(\sqrt{\frac{2x}{\pi}}j_n(x)\right) = J_{n+\frac{1}{2}}(x) \quad (4.4.3)$$

and, after standard manipulations, Eq. (4.4.3) eventually yields the *spherical Bessel equation* in standard form

$$\{x^2\hat{D}_x + 2x\hat{D}_x + [x^2 - n(n+1)]\}j_n(x) = 0. \quad (4.4.4)$$

Regarding the relevant umbral representation, it is sufficient to look at the relevant series expansion

$$j_n(x) = \sqrt{\frac{\pi}{2x}}\sum_{r=0}^{\infty}\frac{(-1)^r}{r!\,\Gamma\left(n+r+\frac{3}{2}\right)}\left(\frac{x}{2}\right)^{n+\frac{1}{2}+2r}, \quad (4.4.5)$$

to end up with

$$j_n(x) = \sqrt{\frac{\pi}{2x}}\left(\frac{\hat{c}x}{2}\right)^{n+\frac{1}{2}}e^{-\hat{c}\left(\frac{x}{2}\right)^2}\varphi_0. \quad (4.4.6)$$

The "umbral technicalities" regarding the spherical Bessel family are the same as the already discussed standard forms and, therefore, we do not experience any difficulty in getting the following results.

Example 38. $\forall x \in \mathbb{R}$

$$\text{(i)} \quad \int_{-\infty}^{\infty}j_0(x)dx = \frac{\sqrt{\pi\hat{c}}}{2}\hat{c}^{\frac{1}{2}}\int_{-\infty}^{\infty}e^{-\hat{c}\left(\frac{x}{2}\right)^2}dx\,\varphi_0$$

$$= \frac{\sqrt{\pi}}{2}\hat{c}^{\frac{1}{2}}\sqrt{\frac{\pi}{\hat{c}}}\varphi_0$$

$$= \frac{\pi}{2}\hat{c}^0\varphi_0 = \frac{\pi}{2}; \quad (4.4.7)$$

$$\text{(ii)} \quad \int_{-\infty}^{\infty}j_0\left(\sqrt{ax^2+bx}\right)dx = \frac{\sqrt{\pi\hat{c}}}{2}\int_{-\infty}^{\infty}e^{-\frac{\hat{c}}{4}(ax^2+bx)}dx\,\varphi_0$$

$$= \frac{\sqrt{\pi\hat{c}}}{2}\sqrt{\frac{\pi}{\hat{c}}}e^{\hat{c}\frac{b^2}{16a}} = \frac{\pi}{2}I_0\left(\frac{b}{2\sqrt{a}}\right). \quad (4.4.8)$$

Equation (4.4.8) can also be viewed as an integral representation of the modified first kind cylindrical Bessel of zeroth order.

Example 39. In analogy to what has been done in Section 4.1, in particular in Eq. (4.1.50), we can easily obtain the following generating function

$$\sum_{n=0}^{\infty} \frac{t^n}{n!} j_n(x) = \frac{\sqrt{\pi}}{2} \hat{c}^{\frac{1}{2}} e^{-\hat{c}\left[\left(\frac{x}{2}\right)^2 - \frac{xt}{2}\right]} \varphi_0 = \sqrt{\frac{\pi}{2}} \frac{J_{\frac{1}{2}}(\sqrt{x^2 - 2xt})}{\sqrt[4]{x^2 - 2xt}}$$

$$= j_0(\sqrt{x^2 - 2xt}). \qquad (4.4.9)$$

Furthermore, noting the operational rule (see Section 7.3)

$$e^{\lambda \frac{1}{z} \partial_z} f(z) = f(\sqrt{z^2 + 2\lambda}), \qquad (4.4.10)$$

we can infer the important property

$$\sum_{n=0}^{\infty} \frac{t^n}{n!} j_n(x) = \sum_{m=0}^{\infty} \frac{(-1)^m}{m!} \xi^m \left(\frac{1}{x} \partial_x\right)^m j_0(x) \mid_{\xi = xt} \qquad (4.4.11)$$

which, along with the identity in Eq. (4.4.9), yields the following (well-known) definition of nth order spherical Bessel in terms of the 0-th

$$j_n(x) = (-x)^n \left[\frac{1}{x} \partial_x\right]^n j_0(x). \qquad (4.4.12)$$

We conclude this section using an umbral representation to write the successive derivatives of the 0th spherical Bessel with respect to the argument.

Example 40. The procedure we apply is the same as to the case relevant to cylindrical Bessel and indeed we get

$$\partial_x^n j_0(x) = \frac{\sqrt{\pi}}{2} \hat{c}^{\frac{1}{2}} \partial_x^n e^{-\hat{c}\left(\frac{x}{2}\right)^2} \varphi_0 \qquad (4.4.13)$$

which, after applying what we have learned about the successive derivative of a Gaussian in terms of two-variable Hermite polynomials, yields

$$\partial_x^n j_0(x) = (-1)^n \frac{\sqrt{\pi}}{2} \hat{c}^{\frac{1}{2}} H_n\left(\frac{\hat{c}}{2}x, -\frac{\hat{c}}{4}\right) e^{-\hat{c}\left(\frac{x}{2}\right)^2} \varphi_0, \qquad (4.4.14)$$

thus eventually finding

$$\partial_x^n j_0(x) = (-1)^n n! \sum_{r=0}^{\lfloor \frac{n}{2} \rfloor} \left(\frac{1}{2}\right)^{n-2r} \frac{(-x)^{-r} j_{n-2r}(x)}{r!(n-2r)!}$$

$$= (-1)^n n! \frac{\sqrt{\pi}}{2} \sum_{r=0}^{\lfloor \frac{n}{2} \rfloor} \frac{(-1)^r \left(\frac{x}{2}\right)^{n-2r}}{2^{2r} r!(n-2r)!} \hat{c}^{n-r+\frac{1}{2}} e^{-\hat{c}\left(\frac{x}{2}\right)^2} \varphi_0$$

$$= (-1)^n n! \sum_{r=0}^{\lfloor \frac{n}{2} \rfloor} \frac{(-1)^r x^{-r} j_{n-r}(x)}{2^r r!(n-2r)!}, \qquad (4.4.15)$$

which, after abusing the umbral notation, can also be written as

$$\partial_x^n j_0(x) = H_n\left(-\hat{j}, -\frac{\hat{j}}{2x}\right) \varkappa_0; \qquad \hat{j}^r \varkappa_0 = j_r(x) \qquad (4.4.16)$$

The study and restyling of Bessel functions, due to the wealth of the relevant properties and forms, is an endless effort. Before concluding this section we like to mention a further *BF* family member that is amenable to a useful umbral translation.

The *Humbert Bessel functions* are an example of two index Bessel, which we will touch up on in the chapter dedicated to exercises and complements. It is defined through the series expansion

$$J_{\mu,\nu}(x) = \sum_{k=0}^{\infty} \frac{(-x)^k}{k!\,\Gamma(k+\mu+1)\Gamma(k+\nu+1)}. \qquad (4.4.17)$$

The importance of this family of functions stems from their use in the theory of electromagnetic processes, associated with the Synchrotron radiation emission by relativistic electrons, moving in bi-periodic magnetic undulators [40, 73]. We will exploit them in a more mathematical context, which allows us to exploit them to define the Strüve (Bessel) functions, a family of functions with important applications in applied Physics including Aerodynamics [152, 169], Magneto Hydrodynamics and Strüve functions [137], Light Diffraction [118].

Strüve and Humbert functions are linked indeed by the Borel-like transform [17]

$$H_\alpha(x) = \left(\frac{x}{2}\right)^{\alpha+1} \int_0^\infty e^{-s} J_{\frac{1}{2}.\alpha+\frac{1}{2}} \left(s\left(\frac{x}{2}\right)^2\right) ds \qquad (4.4.18)$$

whose properties will be further commented on in Chapter 7.

4.4.1   *Interplay between Circular and Bessel Functions*

We have learned that, within the context of umbral calculus, the border within Gaussian, Lorentzian or Bessel functions is very much flexible. Most of the special polynomials can be reduced to Newton binomials or even to monomials. In this concluding section to the umbral "theory" of BF, we make the further effort of looking at the circular functions and at a smooth transition between trigonometric and Bessel functions.

The first step in this direction is a redefinition of the ordinary circular functions in an umbral fashion. We indeed prove that they are a manifestation of the Gauss function if we take the freedom of writing as follows.

Proposition 30. *We consider the umbral operator* (1.3.13) *of Definition 5,* $_{\alpha,\beta}\hat{d}^\kappa \psi_0 = \dfrac{\Gamma(\kappa+1)}{\Gamma(\alpha\kappa+\beta)}$, *for* $\alpha = 2, \beta = 1$, *then we can recast the ordinary cosine,* $\forall x \in \mathbb{R}$, *as*

$$\cos(x) := e^{-2,1\hat{d}\,x^2} \psi_0. \qquad (4.4.19)$$

To improve the writing, we indicate $_{2,1}\hat{d} = \hat{C}$

$$\cos(x) = e^{-\hat{C}x^2} \psi_0. \qquad (4.4.20)$$

Proof. By expanding Eq. (4.4.20) we recover the Taylor series expansion of the cos function

$$e^{-\hat{C}x^2}\psi_0 = \sum_{r=0}^{\infty}\frac{(-1)^r x^{2r}}{r!}\hat{C}^r\psi_0 = \sum_{r=0}^{\infty}\frac{(-1)^r x^{2r}}{(2r)!}. \qquad (4.4.21)$$

$\square$

It is easy to check the consistency of Eq. (4.4.20) with the elementary properties of the trigonometric functions, by keeping indeed the derivative with respect to x as follows.

Corollary 26. $\forall x \in \mathbb{R}$

$$\partial_x e^{-\hat{C}x^2}\psi_0 = -2x\hat{C}e^{-\hat{C}x^2}\psi_0 = -2x\sum_{r=0}^{\infty}(-1)^r\frac{(r+1)!}{(2r+2)!}\frac{x^{2r}}{r!}$$

$$= -\sum_{r=0}^{\infty}(-1)^r\frac{x^{2r+1}}{(2r+1)!} = -\sin(x). \qquad (4.4.22)$$

It is interesting to recover the cyclical law of the successive derivatives of the circular functions by using the present formalism. With this aim, we recall the identity (1.2.7) $\partial_x^n e^{-ax^2} = H_n(-2ax, -a)e^{-ax^2} = (-1)^n H_n(2ax, -a)e^{-ax^2}$ and, by keeping successive derivatives of both sides of Eq. (4.4.20), we find the following result.

Corollary 27.

$$\partial_x^n e^{-\hat{C}x^2}\psi_0 = (-1)^n H_n(2\hat{C}x, -\hat{C})e^{-\hat{C}x^2}\psi_0$$

$$= (-1)^n n!\sum_{r=0}^{\left[\frac{n}{2}\right]}(-1)^r\frac{(2x)^{n-2r}}{(n-2r)!r!}\cos(x; n-r)$$

$$= (-1)^n n!\sum_{r=0}^{\left[\frac{n}{2}\right]}(-1)^r\frac{x}{(n-2r)!\,r!}\frac{j_{n-r-1}(x)}{(2x)^r}$$

$$= \cos\left(x + n\frac{\pi}{2}\right). \qquad (4.4.23)$$

Within the present context, cos and sin functions are the zeroth and first order cases of a more general class of functions so defined.

Definition 18. We define the class of function $\forall x \in \mathbb{R}, \forall n \in \mathbb{N}$

$$\cos(x; n) := \hat{C}^n e^{-\hat{C} x^2} \psi_0 = \sum_{r=0}^{\infty} \frac{(-1)^r}{r!} \frac{(n+r)!}{[2(n+r)]!} x^{2r}. \qquad (4.4.24)$$

They can be identified with the spherical Bessel functions [112] according to the identity

$$\cos(x; n+1) := \frac{j_n(x)}{2^{n+1} x^n}. \qquad (4.4.25)$$

This last result is an interesting yet unexpected outcome of our formalism, indicating how the umbral procedure we have developed offers a natural way of connecting *circular and Bessel-type functions*, through the use of the exponential function.

The previous identity yields a further thread between elementary and higher transcendental functions that will be further stressed in the forthcoming chapter devoted to generalized trigonometric functions.

Chapter 5

Bessel Functions and Umbral Trigonometries

In this chapter, we develop a new point of view to introduce families of functions, which can be identified as *generalizations of the ordinary trigonometric or hyperbolic functions*. They are defined using a procedure based on umbral methods, inspired by the Bessel Calculus of Bochner, Cholewinsky and Haimo [38]. We propose further extensions of the method and of the relevant concepts as well and obtain new families of integral transform allowing the framing of the previous concepts within the context of generalized Borel transforms.

5.1 From Circular to Bessel Function

The umbral formalism represents a tool going deeper into the nature of the functions themselves yielding the concrete idea of the existence of a thread, which links different families of special functions and polynomials. We show that pushing further the level of abstraction we have proposed, a transition from circular to Bessel functions can be envisaged. The protocol we propose is flexible, direct, straightforward and naturally suited for this type of problem. The method offers new possibilities for the introduction of auxiliary polynomials [54] that are all framed, by the use of purely algebraic manipulations, in a context that can be understood as that of generalized trigonometry.

5.1.1 *Umbral Version of the Trigonometric Functions*

In this section, we introduce the Bessel functions by enhancing their "trigonometric" content. The means to accomplish this task is the Umbral formalism.

Observation 8. The differential equation satisfied by the functions (4.4.24) can be derived from those of circular Bessel [2] according to the identities

$$Z_n(x) = \cos(x; n), \quad j_{n-1}(x) = 2^n x^{n-1} Z_n(x),$$

$$x Z_n''(x) + 2n Z_n'(x) + x Z_n(x) = 0. \tag{5.1.1}$$

Regarding the integrals of functions (4.4.24), we find the following result.

Corollary 28. *By the use of GWI (1.2.2), we obtain*

$$\int_{-\infty}^{+\infty} \cos(x; n)\, dx = \int_{-\infty}^{+\infty} \left[\hat{C}^n e^{-\hat{C}x^2} \psi_0 \right] dx = \hat{C}^n \int_{-\infty}^{+\infty} e^{-\hat{C}x^2}\, dx\, \psi_0$$

$$= \sqrt{\frac{\pi}{\hat{C}}}\, \hat{C}^n \psi_0 = \sqrt{\pi}\, \hat{C}^{n-\frac{1}{2}} \psi_0 = \sqrt{\pi} \frac{\Gamma\left(n + \frac{1}{2}\right)}{\Gamma(2n)}. \tag{5.1.2}$$

Further insight into the "genesis" of the trigonometric functions can be obtained by applying again the Gauss–Weierstrass transform method (1.2.2) as follows.

Definition 19. $\forall x \in \mathbb{R}$, we introduce the *Bessel-trigonometric function*

$$c_0^{\left(\frac{1}{2}\right)}(x) = e^{-\hat{C}^{\frac{1}{2}}x} \psi_0 = \sum_{r=0}^{\infty} \frac{(-x)^r}{r!} \hat{C}^{\frac{r}{2}} \psi_0 = \sum_{r=0}^{\infty} \frac{\Gamma\left(\frac{r}{2} + 1\right)}{(r!)^2} (-x)^r. \tag{5.1.3}$$

Proposition 31. $\forall x \in \mathbb{R}$

$$e^{-\hat{C}x^2} \psi_0 = \frac{1}{\sqrt{\pi}} \int_{-\infty}^{+\infty} e^{-\xi^2} \left[e^{-2i\hat{C}^{\frac{1}{2}}x\xi} \psi_0 \right] d\xi$$

$$= \frac{1}{\sqrt{\pi}} \int_{-\infty}^{+\infty} e^{-\xi^2} c_0^{\left(\frac{1}{2}\right)}(2ix\xi)\, d\xi. \tag{5.1.4}$$

We note that, by keeping the successive derivatives of the function defined in Eq. (5.1.3), we can state Proposition 32.

Proposition 32. $\forall p \in \mathbb{N}$

$$\partial_x^p c_{0,0}^{\left(\frac{1}{2}\right)}(x) := \partial_x^p c_0^{\left(\frac{1}{2}\right)}(x) = \partial_x^p \left[e^{-\hat{C}^{\frac{1}{2}}x} \psi_0 \right] = (-1)^p \left[\hat{C}^{\frac{p}{2}} e^{-\hat{C}^{\frac{1}{2}}x} \psi_0 \right]$$

$$= (-1)^p \sum_{r=0}^{\infty} \frac{\Gamma\left(\frac{r+p}{2}+1\right)}{r!\,(r+p)!}(-x)^r, \qquad (5.1.5)$$

which can be associated with the special function

$$c_{\mu,\alpha}^{(\nu)}(x) = \sum_{r=0}^{\infty} \frac{\Gamma\left(\nu\,r + \alpha + 1\right)}{r!\,\Gamma(r + \mu + 1)}(-x)^r, \qquad (5.1.6)$$

and therefore

$$\partial_x^p c_{0,0}^{\left(\frac{1}{2}\right)}(x) = (-1)^p c_{p,\frac{p}{2}}^{\left(\frac{1}{2}\right)}(x). \qquad (5.1.7)$$

Observation 9. The origin of the functions (5.1.6) can easily be traced back to the Tricomi–Bessel functions $C_\beta(x) = \sum_{r=0}^{\infty} \frac{(-x)^r}{r!\,\Gamma(r+\beta+1)}$ shown in Eq. (1.1.2) and are recognized to be associated with an extension of the Borel transform of the functions (5.1.8) (see Section 1.1.1), namely [54]

$$c_{\mu,\alpha}^{(\nu)}(x) = \int_0^{\infty} e^{-s} s^{\alpha} C_\mu(s^\nu x)\,ds. \qquad (5.1.8)$$

We derive, as a straightforward exercise, the associated infinite integrals, by considering two paradigmatic examples. The first is rather artificial and concerns the evaluation of the following integral.

Example 41.

$$\int_{-\infty}^{\infty} e^{-a\,x^2} c_{\mu,\alpha}^{(\nu)}(bx)dx = \int_{-\infty}^{\infty} \left[e^{-a\,x^2 - \hat{\chi}_{\mu,\alpha}^{(\nu)}b\,x} \zeta_0 \right] dx = \sqrt{\frac{\pi}{a}} e^{\frac{b^2}{4a}\left(\hat{\chi}_{\mu,\alpha}^{(\nu)}\right)^2} \zeta_0,$$

$$\left(\hat{\chi}_{\mu,\alpha}^{(\nu)}\right)^r \zeta_0 = \frac{\Gamma\left(\nu\,r + \alpha + 1\right)}{\Gamma(r + \mu + 1)}. \qquad (5.1.9)$$

Accordingly, we eventually get

$$\int_{-\infty}^{\infty} e^{-a\,x^2} c_{\mu,\alpha}^{(\nu)}(b\,x)\,dx = \sqrt{\frac{\pi}{a}} \sum_{r=0}^{\infty} \frac{b^{2r}}{(4a)^r r!} \frac{\Gamma\left(2\,r\,\nu + \alpha + 1\right)}{\Gamma\left(2\,r + \mu + 1\right)}.$$

$$(5.1.10)$$

A further and more familiar example, a naïve consequence of this procedure, is the evaluation of the *Fresnel* integral [2]

$$C(x) = \int_x^{+\infty} \cos\left(\xi^2\right) d\xi, \qquad \forall x \in \mathbb{R}, \qquad (5.1.11)$$

for $x = 0$. The use of the identities (4.4.20) yields

Example 42. By applying a variable change,

$$C(0) = \int_0^{+\infty} \left[e^{-\hat{C} x^4} \psi_0\right] dx = \left(\frac{1}{4} \int_0^\infty e^{-y} y^{\frac{1}{4}-1} dy\right) \hat{C}^{-\frac{1}{4}} \psi_0$$

$$= \frac{1}{4} \frac{\Gamma\left(\frac{1}{4}\right) \Gamma\left(\frac{3}{4}\right)}{\Gamma\left(\frac{1}{2}\right)} = \frac{1}{2}\sqrt{\frac{\pi}{2}}. \qquad (5.1.12)$$

The previous results have emerged in quite a natural fashion from our formalism. The results we have obtained so far can perhaps be obtained with a more conventional formalism. The derivation is however neither easy nor natural.

5.1.2 *Laguerre Polynomials and Trigonometric Function*

In this section, we provide a link betweeen trigonometric-like functions and Laguerre polynomials. Before proceeding in this direction we need the definition of further auxiliary tools. The strategy we follow is that of introducing an appropriate family of polynomials providing a bridge between Laguerre and trigonometric functions. With this aim, we replace in Eq. (2.7.1) $\hat{c}$ with $\hat{C}$, thus defining a new family of polynomials suited for our purposes.

Definition 20. $\forall x, y \in \mathbb{R}, \forall n \in \mathbb{N}$, we introduce $\lambda_n(x, y)$ polynomials as follows:

$$\lambda_n(x, y) := \left[(y - \hat{C} x)^n \psi_0\right] = \sum_{s=0}^n \binom{n}{s} (-1)^s y^{n-s} x^s \hat{C}^s \psi_0$$

$$= \sum_{s=0}^n \frac{(-1)^s s!}{(2s)!} \binom{n}{s} y^{n-s} x^s = n! \sum_{s=0}^n \frac{(-1)^s}{(2s)!(n-s)!} y^{n-s} x^s.$$

$$(5.1.13)$$

They are umbrally equivalent to $L_n(x, y)$.

A straightforward application of our procedure yields the following proposition.

Proposition 33. *The $\lambda_n(x, y)$ generating functions*

$$\sum_{n=0}^{\infty} t^n \lambda_n(x, y) = \sum_{n=0}^{\infty} t^n \left[(y - \hat{C}x)^n \psi_0 \right] = \frac{1}{(1 - yt)\left[1 + \dfrac{\hat{C}xt}{1 - yt} \right]} \psi_0$$

$$= \frac{1}{1 - yt} \left[\sum_{r=0}^{\infty} \left(-\frac{\hat{C}xt}{1 - yt} \right)^r \psi_0 \right] = \frac{1}{1 - yt} \bar{e}_0 \left(\frac{xt}{1 - yt} \right),$$

$$\bar{e}_0(x) = \sum_{r=0}^{\infty} (-1)^r \frac{r!}{(2r)!} x^r, \qquad \forall t \in \mathbb{R} :\mid t \mid < \frac{1}{y} \qquad (5.1.14)$$

and[a] $\forall t \in \mathbb{R}$

$$\sum_{n=0}^{\infty} \frac{t^n}{n!} \lambda_n(x, y) = e^{yt} \left[e^{-x\hat{C}t} \psi_0 \right] = e^{yt} \cos(\sqrt{xt}). \qquad (5.1.15)$$

Furthermore, the $\lambda_n(x, y)$ polynomials (5.1.13) are easily shown to satisfy the following recurrences [52].

Properties 15.

$$\partial_y \lambda_n(x, y) = n\,\lambda_{n-1}(x, y), \qquad \hat{\Delta}\,\lambda_n(x, y) = n\,\lambda_{n-1}(x, y),$$

$$\hat{\Delta} = -4x^{\frac{1}{2}} \partial_x x^{\frac{1}{2}} \partial_x = -2\,(1 + 2x\,\partial_x)\,\partial_x \qquad (5.1.16)$$

which, once combined, yield the partial differential equation

$$\begin{cases} \partial_y \lambda_n(x, y) = \hat{\Delta}\,\lambda_n(x, y) \\ \lambda_n(x, 0) = (-1)^n \dfrac{n!}{(2n)!} x^n. \end{cases} \qquad (5.1.17)$$

---

[a]The generating function (5.1.15) indicates that the $\lambda_n(x, y)$ belong to the family of Appéll polynomials in the y variable [5], this is a characteristic shared with the $L_n(x, y)$.

The Cauchy problem (5.1.17) can be solved by standard means and eventually we find the following solution for $\lambda_n(x, y)$:

$$\lambda_n(x, y) = e^{y\hat{\Delta}}\lambda_n(x, 0), \tag{5.1.18}$$

which can be viewed as an operational definition of the $\lambda_n(x, y)$ polynomials. A little step further also yields the identity

$$e^{y\hat{\Delta}}\bar{e}_0(x) = \frac{1}{1-y}\bar{e}_0\left(\frac{x}{1-y}\right). \tag{5.1.19}$$

Equations (5.1.14)–(5.1.19) are closely symilar to analogous identities satisfied by the Laguerre polynomials [3, 141]. In particular, Eq. (5.1.17) is a kind of heat equation involving the differential operator $\hat{\Delta}$. To complete the analogy with Laguerre polynomial families (2.7.6), we introduce the associated λ-polynomials specified by

Definition 21. We define the associated λ-polynomials as

$$\lambda_n^{(\nu)}(x, y) := \hat{C}^\nu (y - \hat{C}x)^n \psi_0 = \sum_{s=0}^{n}(-1)^s \binom{n}{s} y^{n-s}x^s\hat{C}^{\nu+s}\psi_0$$

$$= n!\sum_{s=0}^{n}(-1)^s\frac{\Gamma(\nu+s+1)}{s!\,(n-s)!\Gamma(2(\nu+s)+1)}\,y^{n-s}x^s. \tag{5.1.20}$$

Corollary 29. *The relevant generating function, by applying Eq. (4.4.24), writes*

$$\sum_{n=0}^{\infty}\frac{t^n}{n!}\lambda_n^{(\nu)}(x, y) = e^{yt}\cos(\sqrt{x\,t};\nu). \tag{5.1.21}$$

Before proceeding further we note that the λ-polynomials are linked to other families of polynomials playing an important role in analysis.

Observation 10. The polynomials

$$h_n(x) = \sum_{k=0}^{n}\binom{n+k}{n-k}(-x)^k, \qquad \forall x \in \mathbb{R}, \forall n \in \mathbb{N}, \tag{5.1.22}$$

are orthogonal polynomials with weight function

$$\rho(x) = \frac{1}{2\pi}\sqrt{\frac{4-x}{x}}, \qquad (5.1.23)$$

and are involved both in the theory of Catalan numbers and of the solution of the *Hausdorff moment problem* [37].

The $h_n(x)$ can be readily written in terms of the integral transform of polynomials $\lambda_n(x, y)$ according to the identity

$$h_n(x) = \frac{1}{n!}\int_0^\infty e^{-\xi}\xi^n \lambda_n(x\,\xi, 1)\, d\xi. \qquad (5.1.24)$$

Before concluding this section, we introduce the following λ-based Bessel functions, defined by the generating function

$$\sum_{n=-\infty}^{+\infty} t^n {}_\lambda J_n(x, y) = e^{\frac{(y-\hat{C}\,x)}{2}\left(t-\frac{1}{t}\right)}\psi_0 = e^{\frac{y}{2}\left(t-\frac{1}{t}\right)}\cos\left(\sqrt{\frac{x}{2}\left(t-\frac{1}{t}\right)}\right), \qquad (5.1.25)$$

which are characterized by the Jacobi–Anger-type identity [3]

$$\sum_{n=-\infty}^{+\infty} e^{in\theta}\left({}_\lambda J_n(x, y)\right) = e^{iy\sin(\theta)}\cos\left(\sqrt{ix\sin(\theta)}\right), \qquad (5.1.26)$$

which can be exploited in problems involving nonlinear oscillations, ruled by differential equations of the type

$$Z\,Z'' + Z'^2 + \frac{Z}{2} = 0. \qquad (5.1.27)$$

Observation 11. By keeping $y = 0$, we find ${}_\lambda J_n(x, 0)$, which can be cast in the following series form:

$${}_\lambda J_n(x, 0) = \sum_{r=0}^{+\infty} \frac{(-1)^{3r+n}(2r+n)!}{r!(r+n)![2(2r+n)]!}\left(\frac{x}{2}\right)^{2r+n}. \qquad (5.1.28)$$

In the case of $n = 0$, abusing our umbral notation and by recalling the rule of the Gaussian successive derivatives, write

$$\partial_x^n {}_\lambda J_0(x, 0) = (-1)^n H_n\left(\hat{c}\,\hat{C}^2\,\frac{x}{2}, -\frac{\hat{c}\,\hat{C}^2}{4}\right) e^{-\hat{c}\left(-\hat{C}\frac{x}{2}\right)^2}\psi_0\varphi_0. \qquad (5.1.29)$$

Bessel and circular umbral operators ($\hat{c}$ and $\hat{C}$) act on the "vacua" (ψ_0 and φ_0).

The last examples, regarding artificial construction of Bessel-type functions, have been aimed at further stressing that, even though complicated in their explicit representation in terms of series, the operational method greatly simplifies the study of the properties of the associated special functions.

5.2   From Laguerre to Airy Forms

The cylindrical Bessels are generalizations of the trigonometric functions, while the associated modified forms are an extension of the relevant hyperbolic counterparts [3]. Such an academic identification is non-particularly deep and might be useful for pedagogical reasons or as a guiding element to study their properties as, e.g., those relevant to the asymptotic forms. We must however underline that Bessel and trigonometric/hyperbolic functions share some resemblances only, but they do not display any full correspondence. The search for functions which are "true" generalizations of the trigonometric (t-) or hyperbolic (h-) forms is however recurrent in the mathematical literature. The attempts in this direction can be ascribed to different strategies, roughly speaking the geometrical [95] and the analytical [90] points of view. The first is based on definitions extending to higher powers of the Pythagorean identity of ordinary trigonometric functions, such a program identifies new trigonometries, with their own geometrical interpretation on elliptic curves and with different numbers playing the role of π [95]. The second invokes the analogy with series expansions, differential equations and the theory of special functions. The generalized $t-h$ functions, defined within these two contexts, are different. In particular, those belonging to the geometric strategy can be recognized as elliptic functions, including Jacobi and Weierstrass forms.

In this section, we develop a systematic procedure within the framework of the analytical point of view. We look for "true" generalizations, in the sense that the functions we define allow a one-to-one mapping onto the properties of the elementary t–h functions, like addition or duplication theorems. With this aim, we exploit the methods provided so far about the understanding of Bessel functions as umbral manifestations of Gauss or exponential functions [18]. These conceptual tools, as well as the ideas developed by Cholewinsky and Reneke [39], provide the elements underlying the formalism of this section, aimed at exploring in depth the identification of trigonometric functions associated with Bessel functions, by getting the proper algebraic environment to establish the relevant properties.

Our starting point is the particular following partial differential equation in which we use the results discussed in Chapter 2.

Example 43. We consider, $\forall x, y \in \mathbb{R}$, the initial conditions problem

$$\begin{cases} {}_l\partial_x F(x,y) = {}_l\partial_y F(x,y) \\ F(x,\,0) = x^{\,n} \\ F(0,y) = y^{\,n}, \end{cases} \qquad (5.2.1)$$

where ${}_l\partial_\xi$ is the l-derivative (2.6.8).
It is easily checked that the solution of Eq. (5.2.1), when $x, y > 0$, can be cast in the form [70]

$$\Lambda_n(x,y) = \sum_{r=0}^{n} \binom{n}{r}^2 x^{\,n-r} y^{\,r}, \qquad (5.2.2)$$

where $\Lambda_n(x,y)$ is an example of hybrid polynomial, introduced in [75].

For reasons which will be clear in the following, we introduce a notation borrowed from [90].

Definition 22. We define the composition rule $\forall x, y \in \mathbb{R}, \forall n \in \mathbb{N}$

$$\Lambda_n(x,y) = \sum_{r=0}^{n} \binom{n}{r}^2 x^{\,n-r} y^{\,r} := (x \oplus_l y)^{\,n} \qquad (5.2.3)$$

the *Laguerre binomial sum* (lbs).

It is evident that such a notion is an extension of the Newton binomial, which can be generated by the action of the shifting exponential operator on an ordinary monomial, namely

$$e^{y\,\partial_x}\,x^{\,n} = \sum_{r=0}^{\infty}\frac{y^{\,r}}{r!}\partial_x^{\,r}x^{\,n} = \sum_{r=0}^{n}\frac{y^{\,r}}{r!}\frac{n!}{(n-r)!}x^{\,n-r} = (x+y)^{\,n}. \quad (5.2.4)$$

Corollary 30. *An analogous rule for the generation of lbs can be achieved by replacing the exponential function with the l-exponential* (7.5.2) $_{l}e(\eta) = \sum_{r=0}^{\infty}\frac{\eta^{\,r}}{(r!)^2}$ *(it is a 0-order Bessel–Tricomi function and satisfies the l-eigenvalue equation* $_{l}\partial_x\left(_{l}e(\lambda\,x)\right) = \lambda\left(_{l}e(\lambda\,x)\right)$ *in* (7.5.4)*) and the ordinary derivative with the l-derivative, satisfying the identity* (2.6.13) $_{l}\partial_{\eta}^{\,n} = \partial_{\eta}^{\,n}\eta^{\,n}\partial_{\eta}^{\,n}$. *Accordingly, we find*

$$_{l}e(y\,_{l}\partial_x)\,x^n = \sum_{r=0}^{\infty}\frac{y^r}{(r!)^2}\partial_x^r x^r \partial_x^r x^n = \sum_{r=0}^{n}\frac{y^r}{(r!)^2}\frac{(n!)^2}{[(n-r)!]^2}x^{n-r}$$

$$= \sum_{r=0}^{n}\binom{n}{r}^2 x^{n-r}y^r = (x \oplus_l y)^n. \quad (5.2.5)$$

According to the previous identities, we can also state the following proposition.

Properties 16.

$$_{l}e(y\,_{l}\partial_x)\,_{l}e(x) = {_{l}e(y)}\,_{l}e(x), \quad _{l}e(y\,_{l}\partial_x)\,_{l}e(x) = {_{l}e(x \oplus_l y)}. \quad (5.2.6)$$

The above identities allow the derivation of the following "semi-group" property of the *l*-exponential

$$_{l}e(y)\,_{l}e(x) = {_{l}e(x \oplus_l y)}. \quad (5.2.7)$$

Definition 23. In full analogy with the ordinary Euler formulae, we introduce the *l*-trigonometric $(l-t)$ functions through the identity

$$_{l}e(i\,x) = {_{l}c(x)} + i\,_{l}s(x), \quad (5.2.8)$$

where $l - t$ *cosine* and *sine* functions are specified by the series[b]

$$_l c(x) = \sum_{r=0}^{\infty} \frac{(-1)^r x^{2r}}{(2\,r)!^2}, \qquad _l s(x) = \sum_{r=0}^{\infty} \frac{(-1)^r x^{2r+1}}{(2\,r+1)!^2}. \qquad (5.2.9)$$

Figure 5.1 provides the plot of $_l s(x)$ vs $_l c(x)$ in a Lissajous-like diagram.

The following Properties 17 have been easily checked in [70].

Properties 17. $\forall \alpha \in \mathbb{R}$, $l - t$ *cosine* and *sine* satisfy the identities

$$_l \partial_x \left[_l c(\alpha\,x) \right] = -\alpha\, _l s(\alpha\,x), \qquad _l \partial_x \left[_l s(\alpha\,x) \right] = \alpha\, _l c(\alpha\,x) \qquad (5.2.10)$$

and therefore the "harmonic" equation

$$\left(_l \partial_x \right)^2 \left[_l c(\alpha\,x) \right] = -\alpha^2\, _l c(\alpha\,x), \qquad \left(_l \partial_x \right)^2 \left[_l s(\alpha\,x) \right] = -\alpha^2\, _l s(\alpha\,x). \qquad (5.2.11)$$

Before providing further results, it is worth noting the following.

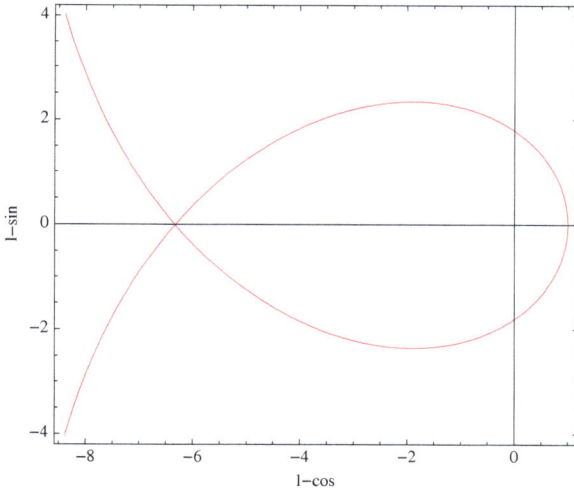

Figure 5.1:   Fish-like Lissajous diagram of l–t functions, $_l s(x)$ vs $_l c(x)$.

---

[b]The $l-h$ functions are defined by the corresponding series expansion
$$_l ch(x) = \sum_{r=0}^{\infty} \frac{x^{2r}}{[(2\,r)!]^2}, \quad _l sh(x) = \sum_{r=0}^{\infty} \frac{x^{2r+1}}{[(2\,r+1)!]^2}.$$

Proposition 34. *Laguerre and ordinary trigonometric functions are linked by the Borel-type transforms* (1.1.18) [70]

$$\int_0^\infty e^{-t} {}_lc(xt)dt = \cos(x), \qquad \int_0^\infty e^{-t} {}_ls(xt)dt = \sin(x). \qquad (5.2.12)$$

The use of the dilatation operator identity (1.1.16)–(1.1.19) *yields the following identifications:*

$${}_lc(x) = (\Gamma(x\,\partial_x + 1))^{-1}\cos(x), \quad {}_ls(x) = (\Gamma(x\,\partial_x + 1))^{-1}\sin(x). \qquad (5.2.13)$$

We note that

$${}_le(i\,x) = {}_lc(x) + i\,{}_ls(x) = (\Gamma(x\,\partial_x + 1))^{-1} e^{i\,x}. \qquad (5.2.14)$$

The use of the property (5.2.7) allows the derivation of the following addition theorems for the functions in Eq. (5.2.9).

Proposition 35. $\forall x, y \in \mathbb{R}$, *the composition rules of* $l-t$ *cosine and sine are stated as*

$$\begin{aligned} {}_lc(x \oplus_l y) &= {}_lc(x){}_lc(y) - {}_ls(x)\,{}_ls(y), \\ {}_ls(x \oplus_l y) &= {}_lc(x){}_ls(y) + {}_ls(x)\,{}_lc(y). \end{aligned} \qquad (5.2.15)$$

Proof. The proof of the second identity is given by noting that

$$\begin{aligned} {}_le(i\,x)\,e(i\,y) &= [{}_lc(x) + i\,{}_ls(x)]\,[{}_lc(y) + i\,{}_ls(y)] \\ &= [{}_lc(x){}_lc(y) - {}_ls(x)\,{}_ls(y)] + i\,[{}_lc(x){}_ls(y) + {}_ls(x)\,{}_lc(y)], \end{aligned} \qquad (5.2.16)$$

and since

$${}_le(i\,x)\,{}_le(i\,y) = {}_le(i\,(x \oplus_l y)) = {}_lc(x \oplus_l y) + i\,{}_ls(x \oplus_l y), \qquad (5.2.17)$$

we can equate real and imaginary parts to infer the identities (5.2.15). The first equation is analogous. □

It is evident that, according to the procedure we have proposed, the properties of ordinary trigonometric functions are extended to their l-counterparts, provided that the ordinary sum is replaced by the composition rule specified in Eq. (5.2.3).

Proposition 36. *The formalism allows the derivation of the corresponding* duplication formulae, *which can be stated by defining the following product rule:*

$$(x \oplus_l x)^n = \frac{(2n)!}{(n!)^2} x^n := (2 \otimes_l x)^n \qquad (5.2.18)$$

which, along with Eq. (5.2.15), yields

$$_lc(2 \otimes_l x) = (_lc(x))^2 - (_ls(x))^2, \qquad _ls(2 \otimes_l x) = 2 \, _lc(x)_ls(x). \quad (5.2.19)$$

Furthermore, the sum can be iterated as

$$(x \oplus_l (x \oplus_l x))^n = (3 \otimes_l x)^n, \qquad (x \oplus_l (\ldots_l (x \oplus_l x)))^k = (n \otimes_l x)^k \qquad (5.2.20)$$

and, accordingly, we can state the following extension of the De Moivre *formulae:*

$$[_lc(x) + i \, _ls(x)]^n = _lc(n \otimes_l x) + i \, _ls(n \otimes_l x). \qquad (5.2.21)$$

It is worth stressing some of the properties of the composition rule (5.2.3).

Properties 18.

(a) $(x \oplus_l y)^n = (y \oplus_l x)^n$,

(b) $(1 \oplus_l 1)^n = \dfrac{(2n)!}{(n!)^2}$,

(c) $(1 \oplus_l (-1))^n = \dfrac{i^n \, n!}{\left(\left(\frac{n}{2}\right)!\right)^2} \dfrac{(1 + (-1)^n)}{2} =$

$$= \begin{cases} 0, & n = 2k + 1, \quad k \in \mathbb{N}, \\ \dfrac{i^n \, n!}{\left(\left(\frac{n}{2}\right)!\right)^2}, & n = 2k, \qquad k \in \mathbb{N}, \end{cases} \qquad (5.2.22)$$

(d) $(i \oplus_l (-i))^n = \dfrac{(-1)^n \, n! \, (1 + (-1)^n)}{\left(\left(\frac{n}{2}\right)!\right)^2 \, 2} =$

$$q = \begin{cases} = 0, & n = 2k + 1, \quad k \in \mathbb{N}, \\ = \dfrac{(-1)^n \, n!}{\left(\left(\frac{n}{2}\right)!\right)^2}, & n = 2k, \qquad k \in \mathbb{N}. \end{cases}$$

It is furthermore worth noting that

$$(e)_l e(x) = \lim_{n\to\infty} \left(1 \oplus_l \left(\frac{x}{n^2}\right)\right)^n, \tag{5.2.23}$$

which provides the quantity

$$_l e := {}_l e(1) = \lim_{n\to\infty} \left(1 \oplus_l \left(\frac{1}{n^2}\right)\right)^n = 2.279585302336067\ldots \tag{5.2.24}$$

which is a kind of Laguerre-Napier number "$_l e$", presumably trascendent.

Observation 12. The previous identities, albeit trivial, are important to appreciate the structural differences with respect to their ordinary counterparts, for example, Eq. (5.2.22) implies that

$$_l e(i\,x)\,_l e(-i\,x) \neq 1 \tag{5.2.25}$$

and therefore that

$$[_l c(x)]^2 + [_l s(x)]^2 \neq 1. \tag{5.2.26}$$

Observation 13. We observe that correspondent results can be obtained by starting from different special functions. If we consider, e.g., the Mittag–Leffler function (1.3.1) for $\beta = 1$, $E_{\alpha,1} = \sum_{r=0}^{\infty} \frac{x^r}{\Gamma(\alpha r + 1)}$, we note that

$$E_{\alpha,1}(x + y) \neq E_{\alpha,1}(x)\,E_{\alpha,1}(y) \tag{5.2.27}$$

and, therefore, to realize the semigroup properties, we extend the Newton binomial as

$$(x \oplus_{ml_\alpha} y)^n := \sum_{r=0}^{n} \binom{n}{r}_\alpha x^{n-r} y^r,$$

$$\binom{n}{r}_\alpha := \frac{\Gamma(\alpha\,n + 1)}{\Gamma(\alpha\,(n - r) + 1)\,\Gamma(\alpha r + 1)}, \tag{5.2.28}$$

which allows the conclusion

$$E_{\alpha,1}(x \oplus_{ml_\alpha} y) = E_{\alpha,1}(x)\,E_{\alpha,1}(y). \tag{5.2.29}$$

The associated sin and cos-like functions defined by

$$C_{\alpha,1}(x) = \frac{E_{\alpha,1}(ix) + E_{\alpha,1}(-ix)}{2}, \quad S_{\alpha,1}(x) = \frac{E_{\alpha,1}(ix) - E_{\alpha,1}(-ix)}{2\,i} \tag{5.2.30}$$

also imply that

$$E_{\alpha,1}(ix) = C_{\alpha,1}(x) + i\,S_{\alpha,1}(x) \tag{5.2.31}$$

and they are characterized by the addition formulae

$$C_{\alpha,1}(x \oplus_{ml_\alpha} y) = C_{\alpha,1}(x)\,C_{\alpha,1}(y) - S_{\alpha,1}(x)\,S_{\alpha,1}(y),$$
$$S_{\alpha,1}(x \oplus_{ml_\alpha} y) = S_{\alpha,1}(x)\,C_{\alpha,1}(y) + C_{\alpha,1}(x)\,S_{\alpha,1}(y), \tag{5.2.32}$$

resembling those of their circular counterpart. It is furthermore worth noting that, as we noted in Lemma (15), if $\alpha = n \in \mathbb{N}$, the ML function satisfies the eigenvalue equation

$$n^n \left(x^{\frac{n-1}{n}} \frac{d}{dx} \right)^n E_{n,1}(\lambda x) = \lambda E_{n,1}(\lambda x). \tag{5.2.33}$$

It is therefore evident that by introducing the ML derivative operator

$$_{ml}\hat{D}_x = n^n \left(x^{1-\frac{1}{n}} \frac{d}{dx} \right)^n, \tag{5.2.34}$$

we find

$$E_{n,1}\left(y \,_{ml}\hat{D}_x \right) E_{n,1}(x) = E_{n,1}(x \oplus_{ml_\alpha} y). \tag{5.2.35}$$

Accordingly, the operator $E_{n,1}\left(y \,_{ml}\hat{D}_x \right)$ is a shift operator in the sense that it provides a shift of the argument of the ML function according to the composition rule established in Eq. (5.2.28).

In the forthcoming sections, we will go deeper into the theory of these families of functions and we will be able to better appreciate the similitudes and the differences with the ordinary forms.

5.2.1 *Generalized Trigonometric Functions, Ordinary and Higher-order Bessel Functions*

The Bessel functions are characterized by a continuous variable and by a real or complex index. A fairly natural extension of the function defined by Eq. (5.2.9) is therefore provided by the l–t functions, associated with the α-order Bessel-like function (7.5.12), for $m = 1$, then

Corollary 31. *We get the function*

$$
{}_{l}e_{\alpha}(\eta) = \sum_{r=0}^{\infty} \frac{\eta^{r}}{r!\,\Gamma(r+\alpha+1)}, \tag{5.2.36}
$$

which is an eigenfunction of the operator [70]

$$
{}_{(\alpha,\,l)}\partial_{x} = \partial_{x}x\,\partial_{x} + \alpha\,\partial_{x} = x^{-\alpha}\partial_{x}x^{\,\alpha+1}\partial_{x}. \tag{5.2.37}
$$

We can now proceed as in the previous section, by noting that the polynomials

$$
\Lambda_{n}(x,y;\alpha) := \sum_{r=0}^{n} \binom{n}{r} \frac{\Gamma(n+\alpha+1)}{\Gamma(n-r+\alpha+1)\,\Gamma(r+\alpha+1)} x^{\,n-r}y^{\,r}
$$
$$
= (x \oplus_{(\alpha,\,l)} y)^{\,n}, \tag{5.2.38}
$$

are solutions of the equation

$$
{}_{(\alpha,\,l)}\partial_{x}l_{n}(x,y;\alpha) = {}_{(\alpha,\,l)}\partial_{y}l_{n}(x,y;\alpha). \tag{5.2.39}
$$

We further define the composition rule

$$
{}_{l}e_{\alpha}(y\,{}_{(\alpha,\,l)}\partial_{x})\,x^{n} = (x \oplus_{(\alpha,\,l)} y)^{n}, \tag{5.2.40}
$$

and state that

$$
{}_{l}e_{\alpha}(y\,{}_{(\alpha,\,l)}\partial_{x})\,{}_{l}e_{\alpha}(x) = {}_{l}e_{\alpha}(x \oplus_{(\alpha,\,l)} y),
$$
$$
{}_{l}e_{\alpha}(y)\,{}_{l}e_{\alpha}(x) = {}_{l}e_{\alpha}(x \oplus_{(\alpha,\,l)} y). \tag{5.2.41}
$$

Finally, by a slight extension of the discussion of the introductory section, we define the $l-t$ functions of α-order as

$$
{}_{(\alpha,\,l)}c(x) = \sum_{r=0}^{\infty} \frac{(-1)^{r}x^{2r}}{[(2\,r)!]\,\Gamma(\alpha + 2r + 1)},
$$
$$
{}_{(\alpha,\,l)}s(x) = \sum_{r=0}^{\infty} \frac{(-1)^{r}x^{2r+1}}{[(2\,r+1)!]\,\Gamma(\alpha + 2r + 2)}, \tag{5.2.42}
$$

which are shown to satisfy the addition theorems (5.2.15) for the composition rule (5.2.40) .

Corollary 32. *The procedure can be further extended by the use of Humbert–Bessel-like functions,[c] which are defined by the series [42]*

$$\iota e_{\alpha,\,\beta}(\eta) = \sum_{r=0}^{\infty} \frac{\eta^r}{r!\,\Gamma(r+\alpha+1)\,\Gamma(r+\beta+1)}. \tag{5.2.43}$$

They satisfy the differential equation

$$[\partial_\eta\,(\alpha + \eta\,\partial_\eta)\,(\beta + \eta\,\partial_\eta)]\,\iota e_{\alpha,\beta}(\eta) = \iota e_{\alpha,\,\beta}(\eta) \tag{5.2.44}$$

and are therefore eigenfunctions of the operator

$$_{(\alpha,\,\beta,l)}\partial_\eta = \partial_\eta\eta\,\partial_\eta\eta\,\partial_\eta + (\alpha+\beta)\,_l\partial_\eta + \alpha\,\beta\,\partial_\eta, \tag{5.2.45}$$

whose r-order derivative is

$$_{(\alpha,\,\beta,l)}\partial_\eta^r x^n = \frac{n!(n+\alpha)!(n+\beta)!}{(n-r)!(n-r+\alpha)!(n-r+\beta)!}x^{n-r}. \tag{5.2.46}$$

By following the same procedure as before, we introduce the composition rule:

$$\left(x \oplus_{(\alpha,\,\beta,\,l)} y\right)^n$$

$$= \sum_{r=0}^{n} \binom{n}{r} \frac{\Gamma(n+\alpha+1)\Gamma(n+\beta+1)x^{\,n-r}y^{\,r}}{\Gamma(n-r+\alpha+1)\Gamma(r+\alpha+1)\Gamma(n-r+\beta+1)\Gamma(r+\beta+1)}, \tag{5.2.47}$$

so that the associated l–t functions, defined as

$$_{(\alpha,\,\beta,l)}c(x) = \sum_{r=0}^{\infty} \frac{(-1)^{\,r}x^{\,2r}}{(2\,r)!\Gamma(\alpha+2r+1)\,\Gamma(\beta+2\,r+1)},$$

$$\tag{5.2.48}$$

$$_{(\alpha,\,\beta,l)}s(x) = \sum_{r=0}^{\infty} \frac{(-1)^{\,r}x^{\,2r+1}}{(2\,r+1)!\Gamma(\alpha+2r+2)\,\Gamma(\beta+2\,r+2)},$$

---

[c]In a slightly different way from the definition of Humbert–Bessel-like functions in (4.4.17), we used here $\iota e_{\alpha,\beta}(\eta) = J_{\alpha,\beta}(-\eta)$.

are straightforwardly shown to satisfy the differential equations

$$(\alpha,\beta,l)\partial_\eta\left[(\alpha,\beta,l)c(\lambda x)\right] = -\lambda\left[(\alpha,\beta,l)s(\lambda x)\right],$$

$$(\alpha,\beta,l)\partial_\eta\left[(\alpha,\beta,l)s(\lambda x)\right] = \lambda\left[(\alpha,\beta,l)c(\lambda x)\right]. \tag{5.2.49}$$

and the addition theorems based on the extension of the definition of sum specified in Eqs. (5.2.6) *and* (5.2.41).

We have shown that the concept of $l-t$ function is a fairly natural consequence of the notion of Laguerre derivative, of its extensions and of the associated eigenfunctions, which belong to Bessel-like forms. In the following, we will show how to frame the Cholewinsky–Reneke $l-h$ functions within the present framework. Before entering into this specific aspect of the problem, we introduce some consequences of the previous formalism on the theory of diffusion equation associated to the Laguerre derivative and to its generalization.

5.2.2 *Bessel Diffusion Equations*

This short section, in which we discuss some evolutive equations based on the operators introduced in the previous sections, is an apparent detour from the main stream of the chapter.

Laguerre-type diffusive equations like [17]

$$\begin{cases} \partial_\tau F(x,\tau) = {}_l\partial_x F(x,y) \\ F(x,0) = f(x), \end{cases} \tag{5.2.50}$$

can be formally solved as

$$F(x,\tau) = e^{\tau\,{}_l\partial_x}f(x). \tag{5.2.51}$$

To make the above solution meaningful, it is necessary to specify how to calculate the action of the exponential operator containing the Laguerre derivative on the function $f(x)$. We discuss, therefore, as introductory example, the case in which $f(x) = e^x$, and proceed as follows:

(1) We note that the exponential can be written as an integral transform of the Tricomi function

$$e^x = \int_0^\infty e^{-t}{}_le(x\,t)\,dt. \tag{5.2.52}$$

(2) We use the properties (5.2.6) to end up with

$$e^{\tau \, {}_\iota \partial_x} e^x = \int_0^\infty e^{-t\,(1-\tau)} {}_\iota e(x\,t)\,dt = \frac{1}{1-\tau} e^{\frac{x}{1-\tau}}. \qquad (5.2.53)$$

More in general, whenever

$$f(x) = \int_0^\infty \tilde{f}(t) {}_\iota e(x\,t)\,dt, \qquad (5.2.54)$$

the solution of the problem (5.2.50) can be cast in the form

$$F(x,\,\tau) = \int_0^\infty \tilde{f}(t)\, e^{t\,\tau} {}_\iota e(x\,t)\,dt \qquad (5.2.55)$$

and $\tilde{f}(t)$ is the ι-transform of the function $f(x)$.

Before going further with the above formalism, we note that the equation

$$\begin{cases} \partial_\tau F(x,\tau) = {}_{(\alpha,\,\beta,\,\iota)}\partial_x F(x,y) \\ F(x,\,0) = f(x) \end{cases} \qquad (5.2.56)$$

can be solved in an analogous way provided that we replace ${}_\iota e(x)$ with ${}_\iota e_{\alpha,\,\beta}(x)$ in Eq. (5.2.55).

In the case in which we consider equations of the type

$$\begin{cases} {}_\iota \partial_\tau F(x,\tau) = {}_\iota \partial_x F(x,y) \\ F(x,\,0) = f(x), \end{cases} \qquad (5.2.57)$$

the solution of the problem can be obtained as

$$F(x,\,\tau) = {}_\iota e(\tau\,{}_\iota \partial_x) f(x) = f(x \oplus_\iota \tau). \qquad (5.2.58)$$

Further comments on the previous statements will be provided in the following section.

5.3 Pseudo-Hyperbolic Functions and Generalized Airy Diffusion Equations

As already stressed, the study of generalized forms of trigonometric and of hyperbolic functions is an old leit motiv in the mathematical literature. On the eve of the 1970s, Ricci introduced [145], a family of *pseudo hyperbolic functions* (PHF), which will be proven of notable importance for the topics we are discussing.

Definition 24. According to [145], the PHF of third order are defined by the series

$$_{[k,3]}e\,(x) = \sum_{r=0}^{\infty} \frac{x^{3\,r+k}}{(3\,r+k)!}, \qquad k = 0,1,2. \qquad (5.3.1)$$

On account of the properties of the cubic roots of the unit [2]

$$\hat{\omega}_p = e^{\frac{2\,i\,p\,\pi}{3}}, \qquad p = 0,1,2, \qquad \hat{\omega}_p^3 = 1, \qquad p = 0,1,2,$$

$$\hat{\omega}_p^2 + \hat{\omega}_p = -1, \qquad p = 1,2, \qquad\qquad (5.3.2)$$

we can state [76] the following.

Definition 25. We define the *Euler-like exponential formulae* $\forall x \in \mathbb{R}$

$$e^{\hat{\omega}\,x} = \sum_{k=0}^{2} \hat{\omega}^k \,_{[k,3]}e(x), \qquad _{[k,3]}e(x) = \frac{1}{3}\sum_{p=0}^{2} \hat{\omega}_p^k \, e^{\hat{\omega}_p x}. \qquad (5.3.3)$$

Proposition 37. *The PHF of third order are eingenfunctions of the cubic operator*

$$(\partial_x)^3 \,_{[k,3]}e(\lambda\,x) = \lambda^3 {}_{[k,3]}e(\lambda\,x) \qquad (5.3.4)$$

and can be exploited to generalize the exponential translation operator as

$$_{[k,3]}\hat{T}(y) = \,_{[k,3]}e(y\,\partial_x). \qquad (5.3.5)$$

Corollary 33. *In the case of $k = 0$, the action of this operator on an ordinary monomial is given by*

$$_{[0,3]}\hat{T}(y)\,x^{3n} = \frac{1}{3}\left[e^{\hat{\omega}_0\,y\,\partial_x} + e^{\hat{\omega}_1\,y\,\partial_x} + e^{\hat{\omega}_2\,y\,\partial_x}\right]x^{3n}$$

$$= \frac{1}{3}\sum_{\alpha=0}^{2}(x + \hat{\omega}_\alpha y)^{3n} = (x \oplus_{[0,3]} y)^{3n}. \qquad (5.3.6)$$

Corollary 34. *By direct use of the series expansion definition of the function $_{[0,3]}e\,(x)$, we end up with*[d]

$$_{[0,3]}e\,(y\,\partial_x)\,x^{3n} = \sum_{r=0}^{\infty} \frac{y^{3r}}{(3r)!}\,\partial_x^{3r}\,x^{3n} = \sum_{r=0}^{n}\binom{3n}{3r}\,y^{3r}\,x^{3(n-r)}$$

$$= (x \oplus_{[0,3]} y)^{3n}. \tag{5.3.7}$$

It is therefore evident that the following further identities can be stated:

$$_{[0,3]}e\,(y\,\partial_x)\,_{[0,3]}e(x) = \,_{[0,3]}e(x \oplus_0 y),$$
$$_{[0,3]}e\,(y\,\partial_x)\,_{[0,3]}e(x) = \,_{[0,3]}e(y)\,_{[0,3]}e(x) \tag{5.3.8}$$

which, once merged, yield

$$_{[0,3]}e\,(x)\,_{[0,3]}e(y) = \,_{[0,3]}e(x \oplus_{[0,3]} y). \tag{5.3.9}$$

In this way, we have obtained a result allowing the introduction of t-h like functions according to the paradigm developed so far.

By a straightforward generalization of the discussion developed in these last sections, we introduce the *generalized h-functions* defined as follows.

Proposition 38. *We define the generalized h-functions*

$$_{[0,3]}ch\,(x) = \frac{1}{2}\left(_{[0,3]}e(x) + \,_{[0,3]}e\,(-x)\right) = \sum_{r=0}^{\infty} \frac{x^{6r}}{(6r)!},$$

$$_{[0,3]}sh\,(x) = \frac{1}{2}\left(_{[0,3]}e(x) - \,_{[0,3]}e\,(-x)\right) = \sum_{r=0}^{\infty} \frac{x^{6r+3}}{(6r+3)!}, \tag{5.3.10}$$

and easily state that the relevant addition theorems read

$$_{[0,3]}ch\,(\alpha \oplus_{[0,3]} \beta) = \,_{[0,3]}ch(\alpha)_{[0,3]}ch(\beta) + \,_{[0,3]}sh(\alpha)_{[0,3]}sh(\beta),$$

$$_{[0,3]}sh\,(\alpha \oplus_{[0,3]} \beta) = \,_{[0,3]}ch(\alpha)_{[0,3]}sh(\beta) + \,_{[0,3]}sh(\alpha)_{[0,3]}ch(\beta). \tag{5.3.11}$$

---

[d]A straightforward consequence of Eq. (5.3.6) is that (see also [39])

$$(1 \oplus_{[0,3]} 1)^{3n} = \frac{1}{3}\left(2^{3n} + \left(1 + e^{\frac{2i\pi}{3}}\right)^{3n} + \left(1 + e^{\frac{4i\pi}{3}}\right)^{3n}\right) = \frac{1}{3}\left(2^{3n} + (-1)^n 2\right)$$

Analogous conclusions can be reached for $_{[k,m]}e(x)$, $k = 0, \ldots, m-1$.

To obtain the link with the topics discussed so far, we consider a particular case of the function defined in Eq. (5.2.43) [70].

Example 44. We consider the function

$$
{}_{l}e_{\alpha-\frac{1}{3},-\frac{2}{3}}\left(\left(\frac{\eta}{3}\right)^{3}\right) = \sum_{r=0}^{\infty} \frac{\eta^{3r}}{3^{3r}r!\,\Gamma\left(r+\alpha+\frac{2}{3}\right)\Gamma\left(r+\frac{1}{3}\right)}. \tag{5.3.12}
$$

The *associated Laguerre derivative* can be written in terms of the operator

$$
{}_{\alpha}\hat{\vartheta} = {}_{(\alpha-\frac{1}{3},-\frac{2}{3},l)}\partial_{\left(\frac{\eta}{3}\right)^{3}} = \partial_{\eta}\eta^{-3\alpha}\partial_{\eta}\eta^{3\alpha}\partial_{\eta}, \tag{5.3.13}
$$

appearing in the *generalized Airy equation*

$$
\partial_{t}F(x,t) = {}_{\alpha}\hat{\vartheta}\,F(x,t), \tag{5.3.14}
$$

studied in [39].
The function in Eq. (5.3.12) is equivalent, apart from an unessential normalizing factor, to the function exploited by Cholewinsky and Reneke [39] to study the solution of Eq. (5.3.14), which is linked, in particular, to the following expression:

$$
G_{\alpha}(\eta) = \Gamma\left(\frac{1}{3}\right)\Gamma\left(\alpha+\frac{2}{3}\right){}_{l}e_{\alpha-\frac{1}{3},-\frac{2}{3}}\left(\left(\frac{\eta}{3}\right)^{3}\right). \tag{5.3.15}
$$

We can make the previous results more transparent by using the identity

$$
\frac{1}{(3\,n)!}\Gamma\left(n+\frac{2}{3}\right) = \frac{\Gamma\left(\frac{1}{3}\right)\Gamma\left(\frac{2}{3}\right)}{3^{3n}n!\,\Gamma\left(n+\frac{1}{3}\right)}, \tag{5.3.16}
$$

which allows to recast Eq. (5.3.15) in the form

$$
G_{\alpha}(\eta) = \frac{1}{B\left(\frac{2}{3},\alpha\right)}\sum_{r=0}^{\infty}\frac{B\left(r+\frac{2}{3},\alpha\right)\eta^{3r}}{(3r)!}, \tag{5.3.17}
$$

where $B(x,y)$ is the Beta-function (1.0.6), and derive the identity

$$
{}_{\alpha}\hat{\vartheta}\,G_{\alpha}(\lambda\,\eta) = \lambda^{3}G_{\alpha}(\lambda\,\eta). \tag{5.3.18}
$$

Finally, the use of the Euler dilatation operator yields the following integral transform defining the function (5.3.15) in terms of the pseudo-hyperbolic function of third order

$$G_\alpha(\eta) = \frac{1}{B\left(\frac{2}{3}, \alpha\right)} \int_0^1 t^{-\frac{1}{3}}(1-t)^{\alpha-1}{}_{[0,3]}e(\eta\, t^{\frac{1}{3}})dt. \qquad (5.3.19)$$

According to the paradigm developed so far, we use the associated translation operator to define the composition rule

$$G_\alpha(y \,_\alpha\hat{\vartheta}^{\frac{1}{3}})\, x^{3n} :=$$

$$:= \sum_{r=0}^n \binom{n}{r} \frac{\Gamma\left(\alpha+\frac{2}{3}\right)\Gamma\left(\frac{1}{3}\right)\Gamma\left(n+\alpha+\frac{2}{3}\right)\Gamma\left(n+\frac{1}{3}\right) y^{3r} x^{3n-3r}}{\Gamma\left(r+\frac{1}{3}\right)\Gamma\left(n-r+\frac{1}{3}\right)\Gamma\left(n-r+\alpha+\frac{2}{3}\right)\Gamma\left(r+\alpha+\frac{2}{3}\right)}$$

$$= \left(x \oplus_{\alpha[0|3]} y\right)^{3n} \qquad (5.3.20)$$

forming the *generalized lbs for the addition theorem of the relevant l−h functions*.

A natural extension of the previously developed formalism allows the introduction of the l/h-functions. Their definitions and properties, apart from some computational complications, do not produce any significant conceptual progress.

In this section, we have seen how a "wise" combination of different methods borrowed from special function theory, operational and umbral calculus and integral transforms opens new interesting possibilities for the introduction and systematic study of new families of trigonometric functions. In addition, the method yields, as byproduct, the opportunity of getting natural solutions of a large family of PDE belonging to the family of generalized heat equation, whose links with the radial heat equation have been touched upon in [90].

We stress two points just touched upon in the previous part of the chapter.

Remark 1. We have noted in Eq. (5.2.23) that the Laguerre exponential can be obtained through a limit procedure analogous to that involving the Napier number. The same method can, e.g., be used to state that

$$J_0(x) = \lim_{n \to \infty} \left(1 \oplus_l \left(-\left(\frac{x}{2n} \right)^2 \right) \right)^n, \qquad (5.3.21)$$

which is recognized as the asymptotic limit of Laguerre polynomials [3].

The second point we want to emphasize is the geometrical interpretation of the $l - t$ functions from the geometrical point of view. Such an interpretation is provided in Fig. 5.1, where we have considered the Lissajous curves plotting l-sin vs l-cos.

In Fig. 5.2, it is evident that the curves are open since no periodic behaviour is envisaged. However, a kind of self-similarity can be noted when the amplitude of the oscillations increase with increasing x.

The last figure provides the explicit correspondence of the l-sinus and l-cosinus along with the relevant "l-angle", intended as the area intercepted by the l-curve and the segment forming the angle in the positive abscissa direction (Fig. 5.3).

These paragraphs have provided the formalism for a wider understanding of concepts associated with the Laguerre derivative, the underlying algebraic structures and possible extensions to new forms of Trigonometry.

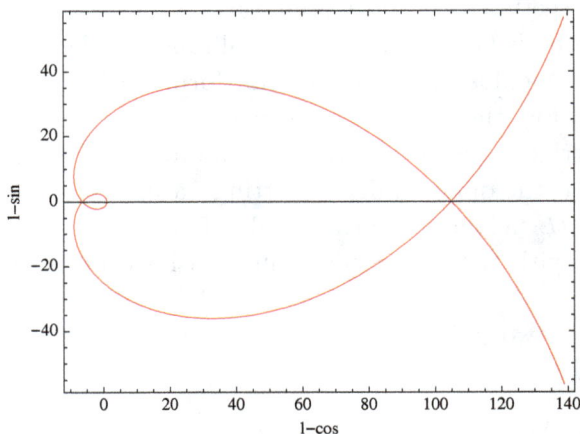

Figure 5.2:   Fish-like Lissajous diagram of l–t functions, $_l s(x)$ vs $_l c(x)$ for larger x-range.

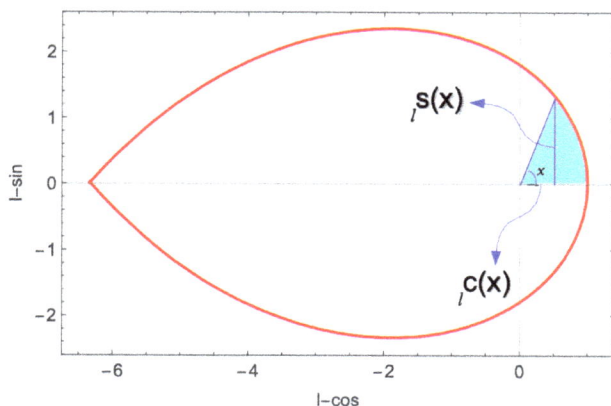

Figure 5.3: *l*-sinus and *l*-cosinus along with parameter x, which is understood as the blue-dashed area.

We have gone through a not fully explored land, the matter we have presented contains elements that are perhaps worth being pursued. We have just provided the elements for autonomous research, for the interested readers.

5.4   Generalized Trigonometric Functions and Matrix Parameterization

Further forms of *Generalized Trigonometric Functions (GTF)* can be introduced by means of an extension of the Euler exponential identity, involving matrices instead of the imaginary unit [69]. The point of view we develop is based on the assumption that matrices are by themselves complex numbers in a broader sense. We see how such an approach yields new families of trigonometry, whose usefulness in application (as charged beam transport) is also discussed. The technique we develop largely benefits from umbral and operational methods, which are the leitmotiv of the treatment we propose.

Definition 26. According to [53, 96], we introduce the GTFs of second order, $C(t)$-$S(t)$, by means of the identity

$$e^{t\,\hat{M}} = C(t)\,\hat{1} + S(t)\,\hat{M}, \qquad \forall t \in \mathbb{R}, \qquad (5.4.1)$$

where $\hat{M}$, $\hat{1}$ are, respectively, a 2×2 non-singular matrix and the unit, namely

$$\hat{M} = \begin{pmatrix} a & b \\ c & d \end{pmatrix}, \qquad \hat{1} = \begin{pmatrix} 1 & 0 \\ 0 & 1 \end{pmatrix}. \qquad (5.4.2)$$

Corollary 35. *From Eq.* (5.4.1), *it also follows that*

$$e^{t\lambda_+} = C(t) + S(t)\lambda_+, \qquad e^{t\lambda_-} = C(t) + S(t)\lambda_-, \qquad (5.4.3)$$

with $\lambda_\pm$ *being the eigenvalues of* $\hat{M}$, *assumed to be non-singular, thus getting the explicit form of the second-order GTF, namely*

$$C(t) = \frac{\lambda_- e^{\lambda_+ t} - \lambda_+ e^{\lambda_- t}}{\lambda_- - \lambda_+}, \qquad S(t) = \frac{e^{\lambda_+ t} - e^{\lambda_- t}}{\lambda_+ - \lambda_-}. \qquad (5.4.4)$$

The structure of Eq. (5.4.1) *is that of the Euler–De Moivre identity, with* $\hat{M}$ *playing the role of an imaginary unit, on the other hand, Eq.* (5.4.3) *represents the scalar counterpart of* (5.4.1) *and, accordingly,* $\lambda_\pm$ *are understood as conjugated imaginary units.*

Properties 19. The properties of the cos and sin like functions $C(t)$, $S(t)$ can be inferred from both Eqs. (5.4.1)–(5.4.3), which yield for example (see also [53])

$$C^2 + \Delta_{\hat{M}} S^2 + Tr(\hat{M})\, CS = e^{Tr(\hat{M})\, t}, \qquad Tr(\hat{M}) = a + d,$$
$$\Delta_{\hat{M}} = a\, d - b\, c, \quad \forall a, b, c, d \in \mathbb{R} : \Delta_{\hat{M}} \neq 0, \qquad (5.4.5)$$

recognized as the fundamental trigonometric identity, and

$$C(2\, t) = C^2 - \Delta_{\hat{M}} S^2, \;\; S(2\, t) = 2\, C(t)\, S(t) + Tr(\hat{M})\, S^2, \qquad (5.4.6)$$

recognized as the duplication formulae.

Properties 20. By keeping the derivative of both sides of Eq. (5.4.1) with respect to the variable t, we find

$$\frac{d}{dt} e^{t\hat{M}} = \left(\frac{d}{dt} C(t)\right)\hat{1} + \left(\frac{d}{dt} S(t)\right)\hat{M}. \qquad (5.4.7)$$

Since

$$\frac{d}{dt} e^{t\hat{M}} = \hat{M} e^{t\hat{M}} = C(t)\hat{M} + S(t)\, \hat{M}^2 \qquad (5.4.8)$$

and since

$$\hat{M}^2 = -\Delta_{\hat{M}} \hat{1} + \mathrm{Tr}(\hat{M})\, \hat{M}, \tag{5.4.9}$$

we end up, after combining Eqs. (5.4.7)–(5.4.9) and equating "real" and "imaginary" parts, the following identities, specifying the properties under derivatives of the *GTF*

$$\frac{d}{dt}C(t) = -\Delta_{\hat{M}} S(t), \qquad \frac{d}{dt}S(t) = Tr(\hat{M})\, S(t) + C(t). \tag{5.4.10}$$

Corollary 36. *We can infer directly from Eq.* (5.4.4) *that the second-order GTFs exhibit, under variable reflection, the identities*

$$C(-t) = e^{-Tr(\hat{M})\,t}\left(-Tr(\hat{M})\,S(t) + C(t)\right) = e^{-Tr(\hat{M})\,t}\left(\frac{d}{dt}S(t)\right),$$

$$S(-t) = -e^{-Tr(\hat{M})\,t} S(t), \tag{5.4.11}$$

which underscore the significant difference with the ordinary TF (be they circular or hyperbolic) with definite even or odd parities.

Further properties can be argued by the use of other means. By keeping, e.g., the freedom of treating $\hat{M}$ as an *ordinary algebraic quantity*, we can formally derive integrals involving *GTF*.

Properties 21.

$$\int^t e^{t'\hat{M}} dt' = {}_I C(t)\, \hat{1} + {}_I S(t)\, \hat{M},$$

$$\int^t e^{t'\hat{M}} dt' = \frac{1}{\hat{M}} e^{t\hat{M}} = C(t)\, \hat{M}^{-1} + S(t)\, \hat{1}, \tag{5.4.12}$$

$${}_I C(t) = \int^t C(t')dt', \qquad {}_I S(t) = \int^t S(t')dt'.$$

Moreover, since the following identity holds:

$$\hat{M}^{-1} = c_{-1}\hat{1} + s_{-1}\hat{M},$$

$$c_{-1} = \frac{\lambda_- \lambda_+^{-1} - \lambda_+ \lambda_-^{-1}}{\lambda_- - \lambda_+} = \frac{Tr(\hat{M})}{\Delta_{\hat{M}}}, \qquad s_{-1} = \frac{\lambda_+^{-1} - \lambda_-^{-1}}{\lambda_+ - \lambda_-} = -\frac{1}{\Delta_{\hat{M}}}, \tag{5.4.13}$$

we obtain the "primitives" of the *GTFs*

$$
{}_IC(t) = \frac{Tr(\hat{M})}{\Delta_{\hat{M}}}C(t) + S(t), \qquad {}_IS(t) = -\frac{1}{\Delta_{\hat{M}}}C(t). \tag{5.4.14}
$$

A straightforward consequence of the previous relationships is reported in the following example.

Example 45.

$$
\int_0^\infty C(-t')dt' = \frac{Tr(\hat{M})}{\Delta_{\hat{M}}}, \qquad \int_0^\infty S(-t')dt' = -\frac{1}{\Delta_{\hat{M}}}, \tag{5.4.15}
$$

which hold true only if the integrals are convergent, namely if $\mathrm{Re}(\lambda_\pm)$ are both positive.

A further slightly more intriguing example is provided by the following *Gaussian* integral.

Example 46.

$$
\int_{-\infty}^{+\infty} e^{-t^2\hat{M}}dt = \sqrt{\frac{\pi}{\hat{M}}} = \sqrt{\pi}\left(c_{-\frac{1}{2}}\hat{1} + s_{-\frac{1}{2}}\hat{M}\right),
$$

$$
c_{-1/2} = \frac{\lambda_-\lambda_+^{-1/2} - \lambda_+\lambda_-^{-1/2}}{\lambda_- - \lambda_+}, \qquad s_{-1/2} = \frac{\lambda_+^{-1/2} - \lambda_-^{-1/2}}{\lambda_+ - \lambda_-}, \tag{5.4.16}
$$

which yields the following generalizations of the *Fresnel integrals*, obtained by other means in Ref. [53],

$$
\int_{-\infty}^{+\infty} C(-t^2)dt = \sqrt{\pi}c_{-\frac{1}{2}}, \qquad \int_{-\infty}^{+\infty} S(-t^2)dt = \sqrt{\pi}s_{-\frac{1}{2}}. \tag{5.4.17}
$$

The convergence of these integrals depends on the eigenvalues $\lambda_\pm$, if convergence is ensured, Eq. (5.4.17) provides the most general form of solution.

This result should be understood in the spirit of our umbral treatment of exponentials and the fact that we treat $\hat{M}$ as an ordinary algebraic quantity.

Corollary 37. *Iterating the procedure, leading to Eqs. (5.4.10), namely by keeping successive derivatives with respect to "t" of both sides of (5.4.1) and by noting that* $\forall n \in \mathbb{N}$

$$\hat{M}^n = c_n \hat{1} + s_n \hat{M}, \tag{5.4.18}$$

we end up with

$$\left(\frac{d}{dt}\right)^n C(t) = c_n C(t) + c_{n+1} S(t),$$
$$\left(\frac{d}{dt}\right)^n S(t) = s_n C(t) + s_{n+1} S(t). \tag{5.4.19}$$

It is evident that the coefficients c_ν, s_ν *are essentially GTF in which* $e^{\lambda_\pm t}$ *are replaced by* $\lambda_\pm^\nu$. *The relevant properties will be discussed later in this chapter.*

Proposition 39. *The* addition formulae *too can be derived in terms of the* c_n, s_n *coefficients as*

$$C(t + t') = C(t)\,C(t') + c_2 S(t)\,S(t'), \qquad c_2 = -\Delta_{\hat{M}},$$
$$S(t + t') = (C(t) + s_2 S(t))\,S(t') + S(t)\,C(t'), \qquad s_2 = \mathrm{Tr}(\hat{M}). \tag{5.4.20}$$

In absence of the simple reflection properties of the ordinary circular functions, we can establish the subtraction formulae *according to the following expressions:*

$$C(t - t') = e^{-s_2 t'} \left[s_2 S(t')C(t) + C(t')C(t) - c_2 S(t)\,S(t') \right],$$
$$S(t - t') = -e^{-s_2 t'} \left[C(t)S(t') - C(t')S(t) \right] \tag{5.4.21}$$

which, once combined with Eq. (5.4.20), yield the following prosthaphaeresis-like *identities:*

$$C(p) - e^{s_2 \frac{(p-q)}{2}} C(q)$$
$$= -s_2 S\left(\frac{p-q}{2}\right) C\left(\frac{p+q}{2}\right) + 2c_2 S\left(\frac{p-q}{2}\right) S\left(\frac{p+q}{2}\right). \tag{5.4.22}$$

In the forthcoming section, we provide some examples aimed at providing the usefulness of this family of functions in applications.

5.5   Evolution Equations Involving Matrices Raised to Non-Integer Exponents

The use of matrices evolution equations raised to non-integer exponents finds applications in problems involving the solution of two- or three-level systems ruled by Klein–Gordon type equations [23]. We develop a fairly simple method exploiting the wealth of results obtained on fractional calculus and provide an example of application.

We discuss the solution of matrix evolution equations, using the formalism of fractional operators, namely of operators raised to a non-integer exponent [69, 71]. We introduce the topics, their scope and the formalism we will employ, by discussing a fairly simple example regarding the solution of the *second-order differential equation*.

Example 47. We consider the differential problem

$$
\begin{cases}
\dfrac{d^2}{dt^2}\,\underline{Y} = -\hat{A}\,\underline{Y} \\[2mm]
\underline{Y}(0) = \underline{Y}_0 \\[2mm]
\dfrac{d}{dt}\,\underline{Y}|_{t=0} = \underline{\dot{Y}}_0,
\end{cases}
\tag{5.5.1}
$$

where $\hat{A}$ is a non-singular 2×2 matrix, with positive defined determinant

$$
\hat{A} = \begin{pmatrix} a & b \\ c & d \end{pmatrix},
\tag{5.5.2}
$$

$\underline{Y} = \begin{pmatrix} y_1 \\ y_2 \end{pmatrix}$ is a two component column vector and $\underline{Y}_0, \underline{\dot{Y}}_0$, the initial conditions of the problem. The solution of Eq. (5.5.1) can be obtained by standard means, namely by introducing a further component $\underline{W} = \frac{d}{dt}\,\underline{Y}$, thus transforming it into a first-order differential equation

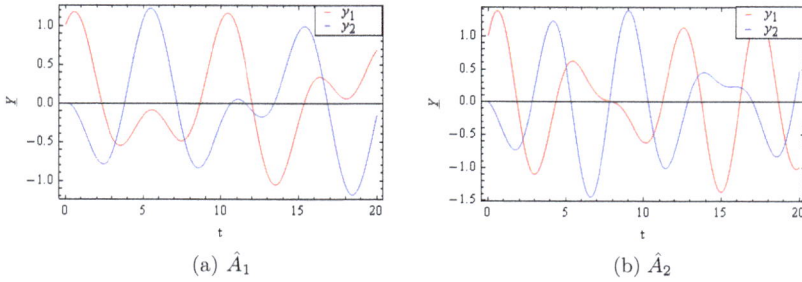

(a) $\hat{A}_1$

(b) $\hat{A}_2$

Figure 5.4: $\underline{Y}$ vs. time; in red and blue the components y_1, y_2, respectively, with initial conditions $\underline{Y}_0 = \begin{pmatrix} 1 \\ 0 \end{pmatrix}$ and $\underline{\dot{Y}}_0 = \begin{pmatrix} -0.1 \\ 0.2 \end{pmatrix}$.

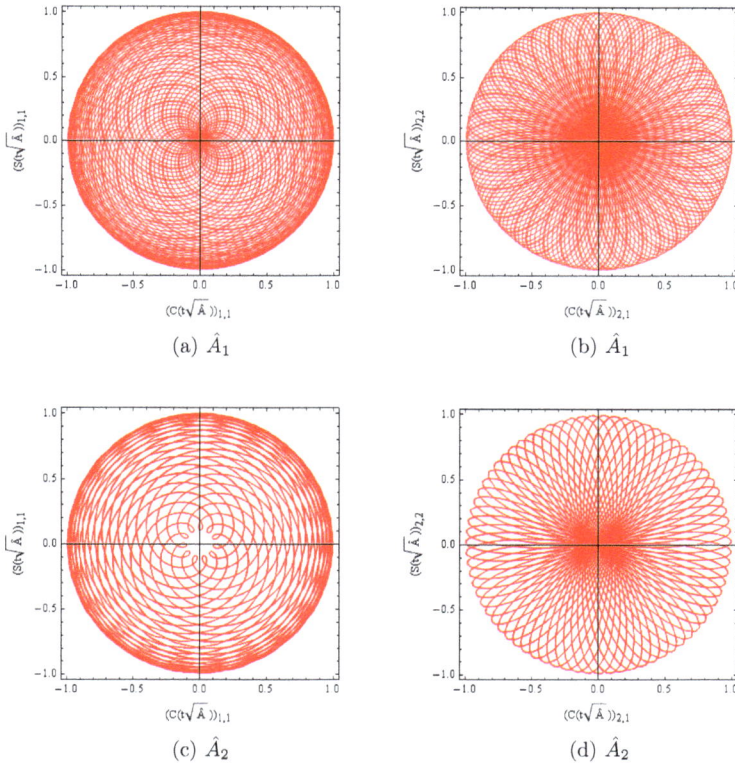

(a) $\hat{A}_1$

(b) $\hat{A}_1$

(c) $\hat{A}_2$

(d) $\hat{A}_2$

Figure 5.5: $S(t\sqrt{\hat{A}})$ vs $C(t\sqrt{\hat{A}})$, with different components combinations for $\hat{A}_1$ (a–b) and $\hat{A}_2$ (c–d).

involving 4×4 matrices. The procedure we follow foresees different means involving the use of square root matrices. Treating the Eq. (5.5.1) as a standard harmonic oscillator equation, we write the relevant solution as

$$\underline{Y}(t) = e^{it\sqrt{\hat{A}}}\underline{c}_1 + e^{-it\sqrt{\hat{A}}}\underline{c}_2, \tag{5.5.3}$$

where $\underline{c}_{1,2}$ are column vectors depending on the initial conditions and will be specified later.

The problem we are faced with is that of providing an operational meaning to an exponential operator containing the square root of a 2×2 matrix. We proceed therefore as follows:

(a) Use standard matrix algebra [27] to write

$$\sqrt{\hat{A}} = -\frac{\lambda_-\lambda_+}{\lambda_- - \lambda_+}\left(\frac{1}{\sqrt{\lambda_-}} - \frac{1}{\sqrt{\lambda_+}}\right)\hat{1} + \frac{\sqrt{\lambda_-} - \sqrt{\lambda_+}}{\lambda_- - \lambda_+}\hat{A},$$

$$f_0(\lambda_+, \lambda_-) = -\frac{\lambda_-\lambda_+}{\lambda_- - \lambda_+}\left(\frac{1}{\sqrt{\lambda_-}} - \frac{1}{\sqrt{\lambda_+}}\right),$$

$$f_1(\lambda_+, \lambda_-) = \frac{\sqrt{\lambda_-} - \sqrt{\lambda_+}}{\lambda_- - \lambda_+}, \tag{5.5.4}$$

with $\hat{1}$, $\lambda_\pm$ being the unit matrix and the eigenvalues of the matrix $\hat{A}$.

(b) Write then

$$e^{\tau\sqrt{\hat{A}}} = e^{f_0(\lambda_+,\lambda_-)\tau\,\hat{1}}e^{f_1(\lambda_+,\lambda_-)\,\tau\,\hat{A}}. \tag{5.5.5}$$

(c) Use the Cayley–Hamilton theorem to write the explicit form of the matrix exponential as (see Refs. [27], [18])

$$e^{\tau\sqrt{\hat{A}}} = e^{f_0(\lambda_+,\lambda_-)\tau}\begin{pmatrix} 1 & 0 \\ 0 & 1 \end{pmatrix}\begin{pmatrix} e_{1,1} & e_{1,2} \\ e_{2,1} & e_{2,2} \end{pmatrix},$$

$$e_{1,1} = \left[\frac{(a-d)\,\tau}{\sqrt{\Delta}}f_1(\lambda_+, \lambda_-)\sinh\left(\frac{\sqrt{\Delta}}{2}\right) + \cosh\left(\frac{\sqrt{\Delta}}{2}\right)\right]$$

$$\cdot e^{\frac{1}{2}f_1(\lambda_+,\lambda_-)\,(a+d)\tau},$$

$$e_{2,2} = \left[-\frac{(a-d)\,\tau}{\sqrt{\Delta}} f_1(\lambda_+, \lambda_-) \sinh\left(\frac{\sqrt{\Delta}}{2}\right) + \cosh\left(\frac{\sqrt{\Delta}}{2}\right) \right]$$

$$\cdot e^{\frac{1}{2} f_1(\lambda_+, \lambda_-)\,(a+d)\tau},$$

$$e_{1,2} = f_1(\lambda_+, \lambda_-) \frac{2\,b\tau}{\sqrt{\Delta}} \sinh\left(\frac{\sqrt{\Delta}}{2}\right)\, e^{\frac{1}{2} f_1(\lambda_+, \lambda_-)\,(a+d)\tau},$$

$$e_{2,1} = f_1(\lambda_+, \lambda_-) \frac{2\,c\tau}{\sqrt{\Delta}} \sinh\left(\frac{\sqrt{\Delta}}{2}\right)\, e^{\frac{1}{2} f_1(\lambda_+, \lambda_-)\,(a+d)\tau},$$

$$\Delta = (f_1(\lambda_+, \lambda_-)\,\tau)^2 \left((a-d)^2 + 4bc\right). \tag{5.5.6}$$

The solution of Eq. (5.5.1) can accordingly be written as

$$\underline{Y}(t) = C\left(t\sqrt{\hat{A}}\right) \underline{Y}_0 + \frac{1}{\sqrt{\hat{A}}} S\left(t\sqrt{\hat{A}}\right) \underline{\dot{Y}}_0,$$

$$C\left(t\sqrt{\hat{A}}\right) = \frac{e^{it\sqrt{\hat{A}}} + e^{-it\sqrt{\hat{A}}}}{2}, \quad S\left(t\sqrt{\hat{A}}\right) = \frac{e^{it\sqrt{\hat{A}}} - e^{-it\sqrt{\hat{A}}}}{2i},$$

$$\tag{5.5.7}$$

where $C\left(t\sqrt{\hat{A}}\right)$, $S\left(t\sqrt{\hat{A}}\right)$ are pseudo-oscillating cos- and *sin*-like solutions.

A graphical example of such a solution is shown in Fig. 5.4, where we have reported the time behavior of the two components $y_{1,2}$ for different values of the matrix entries: (a) $\hat{A}_1 = \begin{pmatrix} 1 & 0.5 \\ 0.5 & 1 \end{pmatrix}$; (b) $\hat{A}_2 = \begin{pmatrix} 1.7 & 0.5 \\ 0.5 & 1.8 \end{pmatrix}$.

For completnes's sake, we have reported in Figs. 5.5 the so-called Lissajous curves arising from plotting $S(t\sqrt{\hat{A}})$ vs $C(t\sqrt{\hat{A}})$.

We will discuss the application of the method to a physical problem later in this chapter and concentrate on further refinement of the relevant mathematical details. In the following sections, we will extend the method including the integral transform technique and the square root to higher-order matrices, we will discuss a relevant application and will comment further on possible developments.

5.5.1 *Fractional Matrix Exponentiation*

The results so far obtained will be further elaborated, including the extension to higher dimensions, but before addressing this specific aspect of the discussion, we complement the previous treatment by using techniques developed within the context of fractional calculus [105] and more specifically with the context of evolution problems regarding the solution of evolution equations, like *Bethe–Salpeter* or other relativistic forms, involving square roots of differential operators [19].

The theory of integro-differential calculus, namely of derivatives and integrals of non-integer order, has received a significant support from methods associated with the Laplace transform and, within such a framework, the use of Lévy transform [142] has been proved to be an effective tool to deal with exponentials with arguments consisting of fractional operators. In this regard, aim we recall that [54]

$$e^{-p^{\frac{1}{2}}} = \int_0^\infty g_{\frac{1}{2}}(\eta) \, e^{-\eta p} d\eta, \qquad g_{\frac{1}{2}}(\eta) = \frac{1}{2\sqrt{\pi \eta^3}} e^{-\frac{1}{4\eta}}, \qquad (5.5.8)$$

with $g_{\frac{1}{2}}(\eta)$ being the Lévy–Smirnov distribution (used, e.g., in [56] and later in Chapter 6).

Definition 27. According to Eq. (5.5.8), we can express the *exponential of a square root matrix* in terms of the integral transform

$$e^{-\tau\sqrt{\hat{A}}} = \int_0^\infty g_{\frac{1}{2}}(\eta) \, e^{-\eta \tau^2 \hat{A}} d\eta. \qquad (5.5.9)$$

The advantage offered by Eq. (5.5.9) is that the exponential inside the integral (5.5.9) depends on the matrix $\hat{A}$ without any further exponentiation, this allows a direct use of the Cayley–Hamilton theorem to solve the problem of getting an explicit expression for the l.h.s. of Eq. (5.5.9). Without considering the general case, we discuss a few interesting examples.

Example 48. We consider the matrix

$$\hat{A} = \hat{i} = \begin{pmatrix} 0 & -1 \\ 1 & 0 \end{pmatrix}, \tag{5.5.10}$$

namely the "unit circular" matrix satisfying the identity

$$\hat{i}^2 = -\hat{1} \tag{5.5.11}$$

and generalizing the Euler identity

$$e^{\hat{i}\vartheta} = \cos(\vartheta)\,\hat{1} + \sin(\vartheta)\,\hat{i}. \tag{5.5.12}$$

The use of Eq. (5.5.6) yields

$$e^{-\tau\sqrt{\hat{i}}} = e^{-\frac{\sqrt{2}}{2}\tau} \begin{pmatrix} \cos\left(\frac{\sqrt{2}}{2}\tau\right) & -\sin\left(\frac{\sqrt{2}}{2}\tau\right) \\ \sin\left(\frac{\sqrt{2}}{2}\tau\right) & \cos\left(\frac{\sqrt{2}}{2}\tau\right) \end{pmatrix}, \tag{5.5.13}$$

representing a kind of *damped matrix rotation*.

Example 49. Using an almost similar argument we find that the square root of the "unit hyperbolic matrix"

$$\hat{h} = \begin{pmatrix} 0 & 1 \\ 1 & 0 \end{pmatrix} \tag{5.5.14}$$

can be written as

$$\sqrt{\hat{h}} = \frac{1}{2}[(1+i)\,\hat{1} + (1-i)\,\hat{h}]. \tag{5.5.15}$$

The solution of equations like

$$\frac{d^2}{d\tau^2}\underline{Y} = \hat{h}\,\underline{Y} \tag{5.5.16}$$

can accordingly be written as

$$\begin{aligned}
\underline{Y}(\tau) = &\begin{pmatrix} \cosh(\tau_+)\cosh(\tau_-) & \sinh(\tau_+)\sinh(\tau_-) \\ \sinh(\tau_+)\sinh(\tau_-) & \cosh(\tau_+)\cosh(\tau_-) \end{pmatrix} \underline{Y}_0 \\
&+ \frac{1}{\sqrt{\hat{h}}}\begin{pmatrix} \sinh(\tau_+)\cosh(\tau_-) & \cosh(\tau_+)\sinh(\tau_-) \\ \cosh(\tau_+)\sinh(\tau_-) & \sinh(\tau_+)\cosh(\tau_-) \end{pmatrix} \underline{\dot{Y}}_0,
\end{aligned}$$

$$\tau_\pm = \frac{1}{2}(1\pm i)\,\tau. \tag{5.5.17}$$

It is evident that the technique we have envisaged can be extended to higher-order matrices, the only problem is that the procedure becomes slightly more cumbersome from the analytical point of view, but it is easily implemented with Mathematica™.

Example 50. In the case in which $\hat{A}$ is a 3×3 matrix, we find

$$e^{\tau \sqrt{\hat{A}}} = e^{f_0(\lambda_1,\lambda_2,\lambda_3)\tau \hat{1}} e^{f_1(\lambda_1,\lambda_2,\lambda_3)\tau \hat{A}} e^{f_2(\lambda_1,\lambda_2,\lambda_3)\tau \hat{A}^2}, \qquad (5.5.18)$$

with $\lambda_{1,2,3}$ being the associated eigenvalues and

$$\begin{pmatrix} f_0 \\ f_1 \\ f_2 \end{pmatrix} = \begin{pmatrix} 1 & \lambda_1 & \lambda_1^2 \\ 1 & \lambda_2 & \lambda_2^2 \\ 1 & \lambda_3 & \lambda_3^2 \end{pmatrix}^{-1} \begin{pmatrix} \sqrt{\lambda_1} \\ \sqrt{\lambda_2} \\ \sqrt{\lambda_3} \end{pmatrix}. \qquad (5.5.19)$$

The explicit form of the matrix can be written in terms of the exponential of the matrix $\hat{A}$ if we use the identity

$$e^{f_2 \tau \hat{A}^2} = \frac{1}{\sqrt{\pi}} \int_{-\infty}^{+\infty} e^{-\xi^2 + 2\xi\sqrt{f_2\tau}\hat{A}} d\xi, \qquad (5.5.20)$$

which yields

$$e^{\tau \sqrt{\hat{A}}} = \frac{e^{\tau f_0}}{\sqrt{\pi}} \begin{pmatrix} 1 & 0 & 0 \\ 0 & 1 & 0 \\ 0 & 0 & 1 \end{pmatrix} \int_{-\infty}^{+\infty} e^{-\xi^2 + (f_1\tau + 2\xi\sqrt{f_2\tau})\hat{A}} d\xi. \qquad (5.5.21)$$

A fairly simple example showing the effectiveness of the method is provided by the matrix

$$\hat{A} = \begin{pmatrix} 0 & -\omega_3 & \omega_2 \\ \omega_3 & 0 & -\omega_1 \\ -\omega_2 & \omega_1 & 0 \end{pmatrix} \qquad (5.5.22)$$

whose exponentiation yields a *Rodrigues matrix* $\hat{R}$ [27, 128]

$$e^{\vartheta \hat{A}} = \&hatR(\vartheta) = \hat{1} + \vartheta \operatorname{sinc}(\Omega \vartheta) \hat{A} + \frac{1}{2}\vartheta^2 \left[\operatorname{sinc}\left(\frac{\Omega \vartheta}{2}\right)\right]^2 \hat{A}^2,$$

$$\Omega = \sqrt{\omega_1^2 + \omega_2^2 + \omega_3^2}. \qquad (5.5.23)$$

The use of the previous identity and a Gaussian integration finally yields the following factorization:

$$e^{\tau \sqrt{\hat{A}}} = e^{\tau f_0} \left[\hat{1} + \left(\frac{1 - e^{-\Omega^2 f_2 \tau}}{\Omega^2} \right) \hat{A}^2 \right] \hat{R}(\tau f_1). \qquad (5.5.24)$$

The extension to higher-order matrices will be discussed in the Chapter 6, containing further comments on the technique we have proposed and some physical applications.

Chapter 6

Number Theory and Umbral Calculus

In this chapter, we provide examples about the use of umbral calculus in Number Theory, in particular on the reformulation of various forms of *Harmonic Numbers* in operatorial form.

6.1 Umbral Methods and Harmonic Numbers

The theory of *Harmonic-based function* is discussed here within the framework of umbral operational methods. We derive a number of results based on elementary notions relying, for example, on the properties of Gaussian integrals. Methods employing the concepts and the formalism of umbral calculus have been exploited in [158] to guess the existence of generating functions involving Harmonic Numbers [156]. The conjectures put forward in [158] have been proven in [43, 46], further elaborated in subsequent papers [126] and generalized to Hyper-Harmonic Numbers in [45].

In this section, we use the same point of view of [158], by discussing the possibility of exploiting the formalism developed therein in a wider context.

6.1.1 *Harmonic Numbers and Generating Functions*

We recall that *Harmonic Numbers* are defined as [156]

$$h_n := \sum_{r=1}^{n} \frac{1}{r}, \qquad \forall n \in \mathbb{N}_0. \tag{6.1.1}$$

It is furthermore evident that the integral representation for this family of numbers can be derived using a standard procedure, reported as follows.

Proposition 40. *In terms of Laplace transform, we obtain*

$$h_n = \sum_{r=1}^{n} \int_0^\infty e^{-sr} ds, \qquad \forall n \in \mathbb{N}_0, \tag{6.1.2}$$

thereby getting nth harmonic number through Euler's integral [115, 148]

$$h_n = \int_0^1 \frac{1 - x^n}{1 - x} dx, \tag{6.1.3}$$

valid more in general $\forall n \in \mathbb{R}^+$.

Proof. $\forall n \in \mathbb{N}_0$, by applying the Laplace transform, the theorem of uniform convergence and the sum of a geometric series, we obtain

$$h_n = \sum_{r=1}^{n} \int_0^\infty e^{-sr} ds = \int_0^\infty \left[\left(\sum_{r=0}^{n} e^{-sr} \right) - 1 \right] ds$$

$$= \int_0^\infty \frac{1 - (e^{-s})^{n+1}}{1 - e^{-s}} - 1 \, ds = \int_{-\infty}^0 \frac{1 - (e^s)^{n+1}}{1 - e^s} - 1 \, ds$$

$$= \int_{-\infty}^0 \frac{e^{(n+1)s} - e^s}{e^s - 1} ds,$$

and by applying the change of variables $e^s \to x$, we obtain

$$h_n = \int_0^1 \frac{1 - x^n}{1 - x} dx.$$

$\square$

According to [115], from this point onwards, the definition in Eq. (6.1.3) can be so extended to *non-natural* values of $\boldsymbol{n}$ and, therefore, it can be exploited as an alternative definition holding for n a *positive real*.

Definition 28. We introduce the function

$$\varphi_h(z) := \varphi_{h_z} = \int_0^1 \frac{1 - x^z}{1 - x} dx, \quad \forall z \in \mathbb{R}^+, \qquad (6.1.4)$$

called *Harmonic number umbral (HNU) vacuum*, or simply vacuum.

Definition 29. The operator

$$\hat{h} := e^{\partial_z} \qquad (6.1.5)$$

is the *vacuum-shift operator*, with z the domain's variable of the function on which the operator acts.

Theorem 4. *The umbral operator,* $\hat{h}^n$*,* $\forall n \in \mathbb{R}^+$ *defines the harmonic numbers,* h_n*, as the action of the shift operator* (6.1.5) *on the HNU vacuum* (6.1.4)

$$\hat{h}^n \varphi_{h_z} \Big|_{z=0} = h_n, \qquad (6.1.6)$$

or simply

$$\hat{h}^n = h_n, \quad h_0 = 0. \qquad (6.1.7)$$

Proof. $\forall n \in \mathbb{R}^+$, by applying the shift operator (6.1.5) on the vacuum (6.1.4), we obtain

$$\hat{h}^n \varphi_{h_0} = \hat{h}^n \varphi_{h_z} \Big|_{z=0} = e^{n\partial_z} \varphi_{h_z} \Big|_{z=0} = \varphi_{h_{z+n}} \Big|_{z=0}$$

$$= \int_0^1 \frac{1 - x^{z+n}}{1 - x} dx \Big|_{z=0} \int_0^1 \frac{1 - x^n}{1 - x} dx = h_n. \quad (6.1.8)$$
$$\square$$

Properties 22. $\forall n, m \in \mathbb{R}^+$, we have

$$\text{(i)} \ \hat{h}^n \hat{h}^m = \hat{h}^{n+m}, \qquad \text{(ii)} \ \left(\hat{h}^n\right)^m = \hat{h}^{n\,m}. \qquad (6.1.9)$$

The proof follows from Eq. (6.1.5).

Definition 30. We call the following series *Harmonic-Based Exponential Function (HBEF)*:

$$_he(x) := e^{\hat{h}\,x} = 1 + \sum_{n=1}^{\infty} \frac{h_n}{n!} x^n, \qquad \forall x \in \mathbb{R}. \qquad (6.1.10)$$

This function, as already discussed in [158], has quite remarkable properties.

The relevant derivatives, $\forall x \in \mathbb{R}, \forall m, k \in \mathbb{N}$, can accordingly be expressed as (see Corollary 3.8 for further comments)

$$\left(\frac{d}{dx}\right)^m {}_h e(x) := {}_h e(x, m) = \hat{h}^m e^{\hat{h} x} = h_m + \sum_{n=1}^{\infty} \frac{h_{n+m}}{n!} x^n,$$

$$(6.1.11)$$

$$\left(\frac{d}{dx}\right)^m {}_h e(x, k) = {}_h e(x, k + m)$$

and, according to Eq. (6.1.10), we also find that

$$\int_0^{\infty} {}_h e(-\alpha x) e^{-x} dx = \int_0^{\infty} e^{-(\alpha \hat{h}+1) x} dx = \frac{1}{\alpha \hat{h} + 1}, \qquad | \alpha | < 1.$$

$$(6.1.12)$$

Corollary 38. *By expanding the umbral function on the r.h.s. of Eq. (6.1.12), we obtain*

$$\frac{1}{\alpha \hat{h} + 1} = 1 + \sum_{n=1}^{\infty}(-1)^n \alpha^n h_n, \qquad | \alpha | < 1.$$

$$(6.1.13)$$

Proof. Using the Taylor expansion and Eq. (6.1.7), $| \alpha | < 1$, we have

$$\frac{1}{\alpha \hat{h} + 1} = \sum_{n=0}^{\infty}(-\alpha \hat{h})^n = 1 + \sum_{n=1}^{\infty}(-1)^n \alpha^n \hat{h}^n = 1 + \sum_{n=1}^{\infty}(-1)^n \alpha^n h_n, \qquad \square$$

which is not an unexpected conclusion, achievable by direct integration, underscored here to stress the consistency of the procedure.
A further interesting example comes from the following "Gaussian" integral.

Example 51.

$$\int_{-\infty}^{\infty} {}_h e(-\alpha x) e^{-x^2} dx = \int_{-\infty}^{\infty} e^{-(\alpha \hat{h} x + x^2)} dx = \sqrt{\pi} e^{\frac{\alpha^2 \hat{h}^2}{4}}, \qquad \forall \alpha \in \mathbb{R}.$$

$$(6.1.14)$$

The last term in Eq. (6.1.14) has been obtained by treating $\hat{h}$ as an ordinary algebraic quantity and then by applying the standard rules of the Gaussian integration (1.2.2).

Observation 14. We note that, using Eq. (6.1.10), we obtain

$$_{h^2}e\left(\frac{\alpha^2}{4}\right) := e^{\frac{\hat{h}^2\alpha^2}{4}} = 1 + \sum_{r=1}^{\infty} \frac{h_{2r}}{r!} \left(\frac{\alpha}{2}\right)^{2r}, \qquad \forall \alpha \in \mathbb{R}. \quad (6.1.15)$$

Let us now consider the following slightly more elaborated example, involving the integration of two "Gaussians", namely the ordinary case and its *HBEF* analogue.

Example 52. $\forall \alpha \in \mathbb{R} : |\alpha| < 1$

$$\int_{-\infty}^{\infty} {_h}e(-\alpha x^2)\, e^{-x^2}\, dx = \int_{-\infty}^{\infty} e^{-(\hat{h}\alpha+1)x^2}\, dx = \sqrt{\frac{\pi}{1+\alpha\hat{h}}}. \quad (6.1.16)$$

This result, obtained after applying elementary rules, can be worded as follows: the integral in Eq. (6.1.16) depends on the operator function on its r.h.s., for which we should provide a computational meaning. The use of the Newton binomial yields

$$\sqrt{\frac{\pi}{1+\alpha\hat{h}}} = \sqrt{\pi}\sum_{r=0}^{\infty}\binom{-\frac{1}{2}}{r}\left(\alpha\hat{h}\right)^r = \sqrt{\pi}\left(1+\sqrt{\pi}\sum_{r=1}^{\infty}\frac{\alpha^r h_r}{\Gamma\left(\frac{1}{2}-r\right)r!}\right). \quad (6.1.17)$$

It is evident that the examples we have provided show that the use of concepts borrowed from umbral theory offers a fairly powerful tool to deal with the "harmonic-based" functions.

6.1.2 *Harmonic-based Functions and Differential Equations*

In the following, we further push the formalism to stress the associated flexibility.

Proposition 41. *The function*

$$\sqrt{h}e(x) := e^{\hat{h}^{\frac{1}{2}}x} = 1 + \sum_{n=1}^{\infty}\frac{\left(\sqrt{\hat{h}}\,x\right)^n}{n!} = 1 + \sum_{n=1}^{\infty}\frac{h_{n/2}}{n!}x^n, \qquad \forall x \in \mathbb{R}, \quad (6.1.18)$$

defines an HBEF, $\forall \alpha \in \mathbb{R}$, through the following Gauss transform:

$$\int_{-\infty}^{+\infty} \sqrt{h}e(\alpha\,x)\,e^{-x^2}\,dx = \int_{-\infty}^{+\infty} e^{\hat{h}^{\frac{1}{2}}\,\alpha\,x - x^2}\,dx = \sqrt{\pi}e^{\hat{h}\left(\frac{\alpha}{2}\right)^2}$$

$$= \sqrt{\pi}\ _h e\left(\left(\frac{\alpha}{2}\right)^2\right). \tag{6.1.19}$$

On the other hand, the function (6.1.18) can be expressed in terms of the $HBEF$, $_h e(x)$, using appropriate integral transform methods [88]. With this aim, indeed, we note that, by the use of the Lévy distribution of order $\frac{1}{2}$: $g_{\frac{1}{2}}(\eta)$ (provided in Eq. (5.5.8)), we can obtain the associated Lévy integral transform

$$e^{-p^{\frac{1}{2}}\,x} = \int_0^\infty e^{-p\,\eta\,x^2}\,g_{\frac{1}{2}}(\eta)\,d\eta, \quad \forall p \in \mathbb{R}^+ \tag{6.1.20}$$

and consequently it allows us to write the following identity.

Corollary 39.

$$\sqrt{h}e(-x) = \int_0^\infty {}_h e(-\eta\,x^2)\,g_{\frac{1}{2}}(\eta)\,d\eta, \qquad g_{\frac{1}{2}}(\eta) = \frac{1}{2\sqrt{\pi\eta^3}}e^{-\frac{1}{4\eta}}. \tag{6.1.21}$$

Proof.

$$\sqrt{h}e(-x) = e^{-\hat{h}^{\frac{1}{2}}x} = \int_0^\infty e^{-\hat{h}\eta x^2}\,g_{\frac{1}{2}}(\eta)\,d\eta = \int_0^\infty {}_h e(-\eta\,x^2)\,g_{\frac{1}{2}}(\eta)\,d\eta. \qquad \square$$

Theorem 5. *The function $_h e(x)$ satisfies the first-order non-homogeneous differential equation*

$$\begin{cases} {}_h e'(x) = \dfrac{d}{dx}{}_h e(x) = {}_h e(x) + \dfrac{e^x - x - 1}{x}, & \forall x \in \mathbb{R} \\ {}_h e(0) = 1. \end{cases} \tag{6.1.22}$$

Proof. Equation (6.1.11), for $m = 1$, yields

$$_h e'(x) := {}_h e(x, 1) = 1 + \sum_{n=1}^\infty \frac{h_{n+1}}{n!}x^n. \tag{6.1.23}$$

Since $h_{n+1} = h_n + \frac{1}{n+1}$, we find

$$_h e(x,1) = 1 + \sum_{n=1}^{\infty} \frac{h_{n+1}}{n!} x^n = {_h e(x)} + \frac{1}{x}\left(e^x - x - 1\right), \qquad (6.1.24)$$

hence, Eq. (6.1.22) follows. □

Corollary 40. *The solution of Eq.* (6.1.22) *yields for HBEF the following explicit expression in terms of ordinary special functions:*

$$_h e(x) = 1 + e^z \left(\ln(x) + E_1(x) + \gamma\right), \qquad E_1(x) = \int_x^{\infty} \frac{e^{-t}}{t}\,dt,$$

$$\left(\ln(x) + E_1(x) + \gamma\right) = -\sum_{n=1}^{\infty} \frac{(-x)^n}{n\,n!},$$

$$(6.1.25)$$

where γ is the Euler–Mascheroni constant.

The previous identity is the generating function of harmonic numbers originally derived by Gosper (see [101, 156]). By iterating the previous procedure, we find the following general recurrence.

Corollary 41.

$$_h e(x,m) = {_h e(x)} + \sum_{r=0}^{m-1} \left(\frac{d}{dx}\right)^r \frac{e^x - 1 - x}{x}. \qquad (6.1.26)$$

Definition 31. The binomial expansion

$$h_n(x) := (x + \hat{h})^n = x^n + \sum_{s=1}^{n} \binom{n}{s} x^{n-s} h_s, \quad \forall x \in \mathbb{R}, \forall n \in \mathbb{N}_0$$

$$(6.1.27)$$

specifies the *Harmonic Polynomials*.

They are easily shown to be linked to the *HBEF* by means of the generating function.

Corollary 42.

$$\sum_{n=0}^{\infty} \frac{t^n}{n!} h_n(x) = e^{xt} {_h e(t)}, \qquad \forall x, t \in \mathbb{R}. \qquad (6.1.28)$$

Proof.

$$\sum_{n=0}^{\infty} \frac{t^n}{n!} h_n(x) = \sum_{n=0}^{\infty} \frac{t^n}{n!}(x+\hat{h})^n = e^{t(x+\hat{h})} = e^{xt}{}_h e(t).$$

□

They belong to the family of Appéll polynomials and satisfy the following recurrences.

Properties 23.

(i) $\dfrac{d}{dx} h_n(x) = n\, h_{n-1}(x), \qquad \forall x \in \mathbb{R},$ \hfill (6.1.29)

(ii) $h_{n+1}(x) = (x+1)\, h_n(x) + f_n(x),$

$$f_n(x) := \sum_{s=1}^{n} \frac{n!}{(n-s)!\,(s+1)!} x^{n-s} = \int_0^1 (x+y)^n dy - x^n, \quad \forall x \in \mathbb{R}.$$

(6.1.30)

Proof. The derivation of Eq. (6.1.29) is trivial. By regarding Eq. (6.1.30), we have:

$$h_{n+1}(x) = (x+\hat{h})(x+\hat{h})^n = (x+\hat{h})\left(x^n + \sum_{s=1}^{n}\binom{n}{s} x^{n-s}\,\hat{h}^s\right)$$

$$= x\, h_n + 1 \cdot x^n + \sum_{s=1}^{n}\binom{n}{s} x^{n-s}\,\hat{h}^{s+1}$$

$$= x\, h_n(x) + \left(x^n + \sum_{s=1}^{n}\binom{n}{s} x^{n-s}\,\hat{h}^s\right) + \sum_{s=1}^{n}\frac{n!\, x^{n-s}}{(n-s)!(s+1)!}$$

$$= (x+1)h_n(x) + \sum_{s=1}^{n}\frac{n!\, x^{n-s}}{(n-s)!(s+1)!},$$

(6.1.31)

and

$$\sum_{s=1}^{n} \frac{n!}{(n-s)!\,(s+1)!} x^{n-s} = \sum_{s=1}^{n} \frac{n!}{s!\,(n-s)!}\frac{x^{n-s}}{s+1} y^{s+1}\Bigg|_{y=1}$$

$$= \sum_{s=1}^{n}\binom{n}{s} x^{n-s}\int_0^1 y^s dy = \int_0^1 \sum_{s=0}^{n}\binom{n}{s} x^{n-s}y^s - x^n dy$$

$$= \int_0^1 (x+y)^n dy - x^n.$$

(6.1.32)

□

Corollary 43. *The identity*

$$h_n(-1) = (-1)^n \left(1 - \frac{1}{n}\right), \qquad \forall n \in \mathbb{N}, \tag{6.1.33}$$

follows from Eq. (6.1.30) after setting $x = -1$. The identity

$$h_n = 1 + \sum_{s=1}^{n} \binom{n}{s} h_s(-1), \qquad \forall n \in \mathbb{N}_0, \tag{6.1.34}$$

is a consequence of the fact that $\hat{h}^n = ((\hat{h} - 1) + 1)^n$.

The harmonic *Hermite* polynomials (touched upon in [43, 158, 180]) can also be written as follows.

Definition 32.

$$\sum_{n=0}^{\infty} \frac{t^n}{n!} \, _hH_n(x) = e^{x\,t} \, _he(t^2), \qquad \forall x, t \in \mathbb{R},$$

$$\tag{6.1.35}$$

$$_hH_n(x) := H_n(x, \hat{h}) = e^{\hat{h}\partial_x^2} x^n = x^n + n! \sum_{r=1}^{\lfloor \frac{n}{2} \rfloor} \frac{x^{n-2\,r}\hat{h}^r}{(n-2\,r)!\,r!}.$$

Properties 24. The recurrence identities of the umbral Hermite polynomials

(i) $\dfrac{d}{dx} \, _hH_n(x) = n \, _hH_{n-1}(x), \qquad \forall x \in \mathbb{R},$

(ii) $_hH_{n+1}(x) = \left(x + 2\,\hat{h}\,\dfrac{d}{dx}\right) \, _hH_n(x)$

$$= \left(x + 2\dfrac{d}{dx}\right) \, _hH_n(x) + 2\,\alpha_n'(x), \tag{6.1.36}$$

$$\alpha_n(x) = n! \sum_{s=1}^{\lfloor \frac{n}{2} \rfloor} \frac{x^{n-2\,s}}{(s+1)!(n-2s)!}, \qquad \alpha_n'(x) = \frac{d}{dx} \alpha_n(x),$$

are a bi-product of the previous identities and a consequence of the monomiality principle in [47].

Corollary 44. *The umbral Hermite satisfies the* second-order non homogeneous ODE

$$\left(x \frac{d}{dx} + 2 \left(\frac{d}{dx} \right)^2 \right) {}_h H_n(x) = n \, {}_h H_n(x) - 2 \, \alpha'_n(x). \qquad (6.1.37)$$

6.1.3   *Truncated Exponential Numbers*

Now, we want to stress the possibility of extending the present procedure to the Truncated Exponential Numbers, namely

$$e_n = \sum_{r=0}^{n} \frac{1}{r!}, \qquad \forall n \in \mathbb{N}. \qquad (6.1.38)$$

The relevant integral representation is [79]

$$e_\alpha := \frac{1}{\Gamma(\alpha+1)} \int_0^\infty e^{-s}(1+s)^\alpha ds, \qquad (6.1.39)$$

which holds for $\alpha \in \mathbb{R}$, too. For example, we provide the following.

Example 53.

$$e_{-\frac{1}{2}} = \frac{e}{\sqrt{\pi}} \Gamma \left(\frac{1}{2}, 1 \right), \qquad (6.1.40)$$

with $\Gamma \left(1, \frac{1}{2} \right)$ being the truncated Gamma function.

According to the previous discussion and Eq. (6.1.40), setting $\hat{e}^\alpha \leftrightarrow e_\alpha$, we also find that

$$\int_{-\infty}^{+\infty} e^{-\hat{e} x^2} dx = \sqrt{\pi} e_{-\frac{1}{2}}, \qquad e^{-\hat{e} x^2} = \sum_{r=0}^{\infty} (-1)^r \frac{e_r}{r!} x^{2r}. \qquad (6.1.41)$$

This last identity is a further proof that the implications offered by the topics treated in this book are fairly interesting and provide countless applications.

6.2   Properties of Generalized Harmonic Numbers

In this section, we introduce *higher-order harmonic* numbers and derive the relevant properties and generating functions by the use of an umbral-type technique.

In [61, 158], different problems concerning harmonic numbers (HN) and the relevant generating functions have been touched upon. The distinctive feature of these investigations is the use of a fairly powerful technique, employing an umbral-like formalism, which has allowed the framing of the theory of HN within an algebraic context. Some of the points raised in [61, 158] have been reconsidered, made rigorous and generalized by means of different technical frameworks in further researches [43, 126].

The following discussion concerns the application of the method foreseen in [61, 158] to generalized forms of harmonic numbers like

$$_m h_n = \sum_{r=1}^{n} \frac{1}{r^m}, \qquad n > 0, \qquad _m h_0 = 0, \qquad (6.2.1)$$

namely *Higher-Order Harmonic Numbers* ($HOHN$) satisfying the property

$$_m h_{n+1} = {}_m h_n + \frac{1}{(n+1)^m}, \qquad (6.2.2)$$

whose associated series provided by the limit $\lim_{n \to \infty} {}_m h_n$, $m > 1$ is, unlike the ordinary HN ($m = 1$), not diverging.

In this section, we derive a number of not previously known properties and the relevant consequences. As an introductory example, we provide the following.

Example 54. We consider the second-order HN ($m = 2$) and write

$$_2 h_n = \int_0^1 \frac{1 - x^n}{x - 1} \ln(x) \, dx, \qquad \forall n \in \mathbb{N}, \qquad (6.2.3)$$

which is obtained after setting

$$\frac{1}{r^2} = \int_0^\infty e^{-sr} s \, ds, \qquad (6.2.4)$$

noting that

$$_2 h_n = \int_0^\infty \frac{e^{-s(n+1)} - e^{-s}}{e^{-s} - 1} s \, ds \qquad (6.2.5)$$

and then by the changing variable of integration.

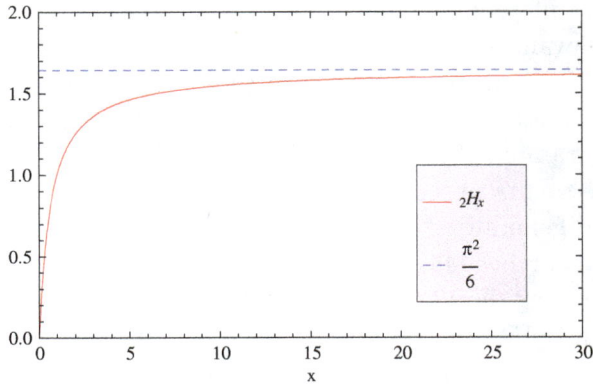

Figure 6.1: $_2h_x$ vs x and $\lim_{x\to\infty} {}_2h_x = \dfrac{\pi^2}{6}$.

It is worth stressing that the integral, representation allows the extension of HN to non-integer indices values. The second-order HN stays between integer and real values of the index, as shown in the plot given in Fig. 6.1, where it is pointed out that the asymptotic limit of the second-order harmonic numbers is $\frac{\pi^2}{6}$.

The relevant extension to negative real indices will be considered later in the chapter.

Let us first consider the generating function associated with the second-order HN, which can be cast in the form of an umbral exponential series according to the following proposition.

Proposition 42. *We introduce*

$$_2h e(t) := 1 + \sum_{n=1}^{\infty} \frac{t^n}{n!} \left({}_2h_n \right) = e^{{}_2\hat{h}t}\xi_0, \qquad (6.2.6)$$

where $_2\hat{h}$ is the umbral operator such that

$$_2\hat{h}^n \xi_0 := \xi_n = {}_2h_n, \quad n > 0, \quad {}_2\hat{h}^0\xi_0 = {}_2h_0 = 1 \qquad (6.2.7)$$

and

$$_2\hat{h}^\nu {}_2\hat{h}^\mu \xi_0 = {}_2\hat{h}^{\nu+\mu}\xi_0, \qquad \forall \nu, \mu \in \mathbb{R}. \qquad (6.2.8)$$

Corollary 45. *From Eqs.* (6.2.5)–(6.2.7), *it follows that*

$$_{2h}e(t, m) = \partial_t^m {_{2h}e(t)} = \partial_t^m e^{2\hat{h}t} = {_2\hat{h}}^m e^{2\hat{h}t} = {_2h_m} + \sum_{n=1}^{\infty} \frac{t^n}{n!} \left(_2h_{n+m}\right).$$

(6.2.9)

Limiting ourselves to the first derivative only, it appears evident that the generating function (6.2.6) *satisfies the identity*

$$\partial_t {_{2h}e(t)} = {_{2h}e(t)} + f_2(t),$$

$$f_2(t) = \sum_{n=1}^{\infty} \frac{t^n}{(n+1)(n+1)!} = \frac{1}{t} \int_0^t \frac{e^{\xi} - \xi - 1}{\xi} \, d\xi = -\frac{Ein(-t) + t}{t},$$

$$Ein(z) = \int_0^z \frac{1 - e^{-\zeta}}{\zeta} d\zeta, \qquad {_{2h}e(0)} = 1.$$

(6.2.10)

Proposition 43. *According to Eq.* (6.2.10), *the problem of specifying the generating function of second-order HN is reduced to the solution of a first-order differential equation, which can be written as*

$$_{2h}e(t) = e^t \left(1 + \sum_{n=1}^{\infty} \frac{1}{(n+1)^2} \left(1 - e^{-t} e_n(t) \right) \right),$$

(6.2.11)

where

$$e_n(x) = \sum_{r=0}^{n} \frac{x^r}{r!}$$

(6.2.12)

are the truncated exponential polynomials [79]. *They belong to the family of Appéll-type polynomials* [5] *and are defined through the operational identity* [101]

$$e_n(x) = \frac{1}{1 - \partial_x} \frac{x^n}{n!}.$$

(6.2.13)

We can further elaborate on the previous identities and set the statement as follows.

Corollary 46.

$$\sum_{n=1}^{\infty} \frac{e_n(t)}{(n+1)^2} = Q_2(t), \qquad Q_2(t) = \frac{1}{1-\partial_t} f_2(t). \qquad (6.2.14)$$

Furthermore, since

$$\sum_{n=1}^{\infty} \frac{1}{(n+1)^2} = \frac{\pi^2}{6} - 1, \qquad (6.2.15)$$

we end up with

$$_2{}_h e(t) = e^t \ \Sigma_2(t), \qquad \Sigma_2(t) = \frac{\pi^2}{6} - Q_2(t) \, e^{-t}. \qquad (6.2.16)$$

This new result can be viewed as an extension of the generating function for the first-order HN derived by Gosper (see what follows) [154].

It is furthermore evident that the formalism allows the straightforward derivation of other identities like in the following.

Lemma 10.

$$\sum_{n=1}^{\infty} \frac{t^n}{n!} \, (_2 h_{n+m}) = e^t \sum_{s=0}^{m} \binom{m}{s} \Sigma_2^{(s)}(t) - {}_2 h_m, \qquad (6.2.17)$$

where the upper index (s) denotes s-order derivative and is a direct consequence of the identity in Eq. (6.2.9).

The extension to $HOHN$, with $m > 2$, follows the same logical steps, namely derivation of the associated Cauchy problem.

Corollary 47. $\forall t \in \mathbb{R}, \forall p \in \mathbb{N} : p > 1$

$$\begin{cases} \partial_t \left({}_{ph} e(t) \right) = {}_{ph} e(t) + f_p(t) \\ f_p(t) = \sum_{n=1}^{\infty} \frac{t^n}{(n+1)^{p-1} \, (n+1)!} \\ {}_{ph} e(0) = 1. \end{cases} \qquad (6.2.18)$$

The Cauchy problem solution can be written as

$$_{ph}e(t) = e^t \left(1 + \sum_{n=1}^{\infty} \frac{1}{(n+1)^p} \left(1 - e^{-t} e_n(t) \right) \right) \qquad (6.2.19)$$

or

$$_{ph}e(t) = e^t \, \Sigma_p(t), \qquad \Sigma_p(t) = \zeta(p) - Q_p(t) e^{-t},$$

$$\zeta(p) = \sum_{n=1}^{\infty} \frac{1}{n^p}, \qquad Q_p(t) = \sum_{n=1}^{\infty} \frac{1}{(n+1)^p} e_n(t) = \frac{1}{1-\partial_t} f_p(t),$$

$$f_p(t) = \sum_{n=1}^{\infty} \frac{t^n}{(n+1)^{p-1} (n+1)!}. \qquad (6.2.20)$$

The case $p = 1$ should be treated separately, because the sum on the r.h.s. of Eq. (6.2.19) is apparently diverging.

It is accordingly worth noting what has been reported in the following observation.

Observation 15.

$$f_1(t) = \sum_{n=1}^{\infty} \frac{t^n}{(n+1)!} = \frac{1}{t}(e^t - t - 1), \qquad (6.2.21)$$

we find

$$_{1h}e(t) = e^t \left(1 + \int_0^t \frac{1 - (\tau + 1) e^{-\tau}}{\tau} d\tau \right) = e^t \, \Sigma_1(t), \qquad (6.2.22)$$

$$\Sigma_1(t) = e^{-t} + Ein(t).$$

Equation (6.2.22) is a restatement of the Gosper derivation of the generating function of first-order HN.

We conclude this paragraph by introducing the following.

Definition 33. We introduce HOHN umbral polynomials (for $m = 1$ see [61])

$$_m H_n(x) = (x + {}_m\hat{h})^n = 1 + \sum_{s=1}^{n} \binom{n}{s} x^{n-s} {}_m H_s, \qquad {}_m H_0(x) = 1.$$

$$(6.2.23)$$

6.3   Motzkin Numbers: an Operational Point of View

The *Motzkin Numbers* are the coefficients of hybrid polynomials. Such an identification allows the derivation of new identities for this family of numbers and offers a tool to investigate previously unnoticed links with the theory of special functions and with the relevant treatment in terms of operational means. The use of umbral methods opens new directions for further developments and generalizations, which leads, e.g., to the identification of new Motzkin associated forms [10].

A very well-known example of a link between special numbers and special polynomials is provided by the so-called convolution or Telephone numbers, which can be expressed in terms of Hermite polynomial coefficients [146]. In [8], the Padovan and Perrin numbers [175, 177] can be recognized to be associated with particular values of two-variable Legendre polynomials [7].

As already underscored, Motzkin numbers [174] have been also discussed in connection with a family of hybrid polynomials [28, 75] and the relevant properties have accordingly been studied.

The hybrid polynomials are indeed defined as [75]

$$P_n^{(q)}(x,y) = n! \sum_{r=0}^{\lfloor \frac{n}{2} \rfloor} \frac{x^{n-2r} y^r}{(n-2r)! \, r! \, (r+q)!}, \qquad (6.3.1)$$

and the relevant generating function reads $\forall t \in \mathbb{R}$

$$\sum_{n=0}^{\infty} \frac{t^n}{n!} P_n^{(q)}(x,y) = \frac{I_q(2\sqrt{y}\,t)}{(\sqrt{y}\,t)^q} e^{xt}, \qquad (6.3.2)$$

where $I_q(x)$ is the modified Bessel function of the first kind of order q (4.3.1).

Within the present framework, the Motzkin number sequence can be specified as [28]

$$m_n = P_n^{(1)}(1,1) = \sum_{s=0}^{n} m_{n,s}, \qquad m_{n,s} = \binom{n}{s} f_s,$$

$$f_s = \frac{s!}{\Gamma\left(\frac{s}{2}+2\right)\Gamma\left(\frac{s}{2}+1\right)} \left|\cos\left(s\frac{\pi}{2}\right)\right|, \qquad (6.3.3)$$

Table 6.1: Motzkin numbers and their coefficients.

Parameter		$m_{n,s}$ coefficients								m_n Motzkin
		s								$\sum_{s=0}^{n} m_{n,s}$
		0	1	2	3	4	5	6	7	
	0	1								1
	1	1	0							1
	2	1	0	1						2
	3	1	0	3	0					4
n	4	1	0	6	0	2				9
	5	1	0	10	0	10	0			21
	6	1	0	15	0	30	0	5		51
	7	1	0	21	0	70	0	35	0	127
	...	...	...	...						...

where the coefficients $m_{n,s}$ can be represented as the triangle reported as follows
in which $m_{n,2}$ corresponds, in OEIS, to the sequence $A000217$, $m_{n,4}$ to $A034827$, $m_{n,6}$ to $A000910$ and so on.

According to Eq. (6.3.2), the Motzkin numbers can also be defined as the coefficients of the following series expansion:

$$\sum_{n=0}^{\infty} \frac{t^n}{n!} m_n = \frac{I_1(2t)}{t} e^t. \qquad (6.3.4)$$

In the following, we show how some progress in the study of the relevant properties can be done by the use of a formalism of umbral nature.

6.3.1 *Motzkin Numbers and Umbral Calculus*

In order to simplify most of the algebra associated with the study of the properties of the Motzkin numbers and to get new relevant identities, we apply our methods of umbral nature.

With this aim we recall umbral form of ν-order Tricomi–Bessel function (3.1.40), $C_\nu(x) = \hat{c}^\nu e^{-\hat{c}x}\varphi_0$. We note that by the use of Eqs. (3.1.41) $C_\nu(x) = \left(\frac{1}{x}\right)^{\frac{\nu}{2}} J_\nu(2\sqrt{x})$ and (4.3.2) $I_\nu(ix) = e^{i\nu\frac{\pi}{2}} J_\nu(x)$, we can write $\forall q \in \mathbb{Z}$ the identity

$$C_q(-x) = \frac{I_q(2\sqrt{x})}{(\sqrt{x})^q} = \sum_{r=0}^{\infty} \frac{x^r}{r!(q+r)!}. \tag{6.3.5}$$

Lemma 11. *The use of this formalism allows to restyle the hybrid polynomials in the form*

$$P_n^{(q)}(x,y) = \hat{c}^q H_n(x, \hat{c}\, y), \tag{6.3.6}$$

where $H_n(x,y)$ are the two-variable HP.

We can accordingly use the wealth of properties of this family of polynomials to derive further and new relations regarding those of the Motzkin numbers family, e.g., see the following proposition.

Proposition 44. *By recalling the generating function (2.4.15)* $\sum_{n=0}^{\infty} \frac{t^n}{n!} H_{n+l}(x,y) = H_l(x + 2yt, y)e^{xt+yt^2}$, *we find*

$$\sum_{n=0}^{\infty} \frac{t^n}{n!} m_{n+l} = \hat{c}\, H_l(1 + 2\hat{c}t, \hat{c})e^{t+\hat{c}t^2}, \tag{6.3.7}$$

which, after using Eq. (3.1.41), finally yields

$$\sum_{n=0}^{\infty} \frac{t^n}{n!} m_{n+l} = \mu_l(t)\, e^t,$$

$$\mu_l(t) = l! \sum_{r=0}^{\lfloor \frac{l}{2} \rfloor} \frac{1}{r!} \sum_{s=0}^{l-2r} \frac{2^s}{s!(l-2r-s)!} \frac{I_{s+r+1}(2t)}{t^{r+1}}. \tag{6.3.8}$$

Furthermore, the same procedure and the use of the Hermite polynomials duplication formula [3]

$$H_{2n}(x,y) = \sum_{r=0}^{n} \binom{n}{r}^2 r!\, (2y)^r \left(H_{n-r}(x,y) \right)^2, \tag{6.3.9}$$

yields the following identity for Motzkin numbers.

Proposition 45.

$$m_{2n} = \hat{c} \sum_{r=0}^{n} r! \binom{n}{r}^2 (2\hat{c})^r H_{n-r}(1,\hat{c}) H_{n-r}(1,\hat{c})$$

$$= \sum_{r=0}^{n} \binom{n}{r}^2 2^r r!(n-r)! \sum_{s=0}^{\lfloor \frac{n-r}{2} \rfloor} \frac{m_{n-r}^{(r+s+1)}}{(n-r-2s)!s!}, \quad (6.3.10)$$

where

$$m_n^{(q)} = P_n^{(q)}(1,1) = \hat{c}^q H_n(1,\hat{c}) \quad (6.3.11)$$

are associated Motzkin numbers [28].

Corollary 48. *The identification of Motzkin numbers as in Eq. (6.3.11), along with the use of the recurrences of Hermite polynomials, yields, e.g., the identities*

$$m_{n+1}^{(q)} = m_n^{(q)} + 2\,n\,m_{n-1}^{(q+1)},$$

$$m_{n+p} = \sum_{s=0}^{\min[n,p]} 2^s s! \binom{p}{s} \binom{n}{s} M_{p-s,\,n-s,\,s},$$

$$M_{p,\,n,\,t} = p! \sum_{r=0}^{\lfloor \frac{p}{2} \rfloor} \frac{m_n^{(t+r+1)}}{(p-2r)!r!}, \quad (6.3.12)$$

in which, the second identity, has been derived from the Nielsen *formula for $H_{n+m}(x,y)$ [131].*

6.3.2 *Telephone Numbers*

In this last part, we have shown that a fairly straightforward extension of the formalism put forward in [28] allows non-trivial progress in the theory of Motzkin numbers. Further relations can be easily obtained by applying the method we have envisaged as, e.g., in Lemma 12.

Lemma 12.

$$\sum_{s=0}^{n} m_{n-s}\, m_s = 2\, (n+1)\, m_n^{(2)} \qquad (6.3.13)$$

is a discrete self-convolution of Motzkin numbers.

We have also mentioned the existence of the associated Motzkin numbers

$$m_n^{(q)} = P_n^{(q)}(1,1), \qquad (6.3.14)$$

touched upon in [28]. In the present context they have been introduced on purely algebraic grounds. Strictly speaking they are not integers and therefore they are not amenable for a combinatorial interpretation. However, redefining them through the following lemmas

Lemma 13. *We recast the associated Motzkin numbers as*

$$\tilde{m}_n^{(q)} = \frac{(n+q)!}{n!} P_n^{(q)}(1,1). \qquad (6.3.15)$$

we obtain for $q = 2$ the sequences in OEIS ($A014531$), while for $q = 3$, the sequences ($A014532$) and so on.

We have mentioned in Section 6.3 the theory of *Telephone numbers* $T(n)$ [113], whose importance in chemical Graph theory has been recently emphasized in [103]. As is well known, they can be expressed in terms of ordinary Hermite polynomials, however, the use of the two-variable extension is more effective. They can indeed be expressed as $T(n) = H_n(1, \frac{1}{2})$.

Lemma 14. *The use of Hermite polynomial's properties, like the index duplication formula, yields*

$$T(2n) = \sum_{r=0}^{n} r! \binom{n}{r}^2 T(n-r)^2. \qquad (6.3.16)$$

Table 6.2: Telephone number coefficients.

Parameter		$t_{n,s}$ **coefficients**							
		s							
		0	**1**	**2**	**3**	**4**	**5**	**6**	**7**
n	**0**	1							
	1	1	0						
	2	1	0	1					
	3	1	0	3	0				
	4	1	0	6	0	3			
	5	1	0	10	0	15	0		
	6	1	0	15	0	45	0	15	
	7	1	0	21	0	105	0	105	0
	...	...	...	...					

Proposition 46. *The use of the Hermite numbers h_s [64] allows the derivation of the following further expression:*

$$T(n) = \sum_{s=0}^{n} t_{n,s}, \qquad t_{n,s} = \binom{n}{s} h_s \left(\frac{1}{2}\right),$$

$$h_s(y) = y^{\frac{s}{2}} \Gamma \left(\frac{s}{2} + 2\right) f_s = \frac{y^{\frac{s}{2}} s!}{\Gamma\left(\frac{s}{2} + 1\right)} \left| \cos\left(s \frac{\pi}{2}\right) \right|. \tag{6.3.17}$$

The coefficients $t_{n,s}$ of the telephone numbers can be arranged as in the triangle in Table 6.2 in which the numbers belonging to the column $s = 4$ $(3, 15, 45, 105, 210, \dots)$ are identified with OEIS A050534, and $s = 5$ $(15, 105, 420, 1260, 3150, \dots)$ is just a multiple of A00910. The use of the identification with two-variable Hermite polynomials opens further perspectives. By exploiting indeed the higher-order HP (see Section 7.4), we can introduce the following proposition.

Proposition 47. *We provide a generalization of telephone numbers*

$$T_n^{(m)} = H_n^{(m)} \left(1, \frac{1}{m}\right), \tag{6.3.18}$$

with generating function

$$\sum_{n=0}^{\infty} \frac{t^n}{n!} T_n^{(m)} = e^{t + \frac{1}{m} t^m}, \tag{6.3.19}$$

which satisfy the recurrence

$$T_{n+1}^{(m)} = T_n^{(m)} + \frac{n!}{(n-m+1)!} T_{n-m}^{(m)}. \qquad (6.3.20)$$

In the case of $m = 3$, we can identify the numbers $T_n^{(3)} = (1, 1, 1, 3, 9, 21, 81, 351, 1233, \ldots)$ with OEIS $A001470$, while for $m = 4$, the series $1, 1, 1, 1, 7, 31, 91, 211, 1681, 12097$ corresponds to $A118934$. For $m = 5$, the associated series appears to be $A052501$ but should be more appropriately identified with the coefficients of the expansion (6.3.18), finally the sequence $n = 6$ is not reported in OEIS.

The discussion of the use of umbral/operatorial methods to study the properties of "special numbers" has been necessarily short. We hope that we have been able to convey the relevant usefulness and that this last chapter has provided a more general view to the possibilities offered by these techniques. The next chapter, containing exercises and complements, has been designed to fill some gaps left open in the previous discussion.

Chapter 7

Complements and Exercises

In this chapter, we discuss a number of exercises and complements, aimed at coroborating the notions acquired in the previous parts of this book. We propose a "gentle" implementation of the already discussed theoretical details, conceived in such a way that the solution strategies are introduced in complementary form to the matter exposed in the main body of the previous chapters and also included in the form of pragmatic instructions. The goal is that of providing a kit-tool to afford "apparently" complicated problems, which find a fairly natural solution, within the context of the umbral formalism and also displaying the link with other important technicalities treated in the book like the Complex Analysis, the Theory of Operators, the Generalized Transforms, Trigonometry and so on. We start with an umbral "revisitation" of very well-known notions in the Theory of Special and Elementary Functions.

7.1 Euler Gamma, Beta Functions and Related Topics

A deeper understanding of the Gamma and Beta functions and of the underlying technicalities are essential elements to manage the practical aspects of the calculus we have outlined in this book. This is however only an aspect of the message we like to convey, we will indeed see how an appropriate combination of Euler functions with the theory of Laplace transform yields a non-trivial complement of the matter we discussed.

We have underscored that the umbral methods realize a simplification of the theory of higher transcendental functions by downgrading the associated degree of transcendence. We have accordingly explored the consequences, which can be drawn by treating a Gaussian as a rational function of the Lorentz type. The emerging features are interesting and open new avenues, allowing the transitions to, e.g., generalized forms of Trigonometric functions and of dispersion relations. Apart from these results, also addressed to stimulate more elaborated thoughts, hopefully inspiring the reader to deeper progress, we believe that the procedure allows an effective understanding of the structure of the function itself displaying the intimate connections with other functions, including the Mittag–Leffler families.

A further effort in this chapter is the merging of the first two topics to study different problems involving integral transform, operational methods and special functions. The associated technicalities and the underlying concepts are by no means a secondary issue, within the context of the umbral methods we are developing in these lectures, and therefore a careful reading is recommended.

The forthcoming exercises, integrations and complements are developed following the steps developed in the main body of the chapter.[a] We cover some points which have been just touched upon or left without adequate comments. The solutions of the proposed exercises are given in the form of hints or as extended and detailed comments, according to the difficulty of the exercise.

1 Consider the integral representation of Euler Γ-function given in Eq. (1.0.1) and use the generating function method (see, e.g., Eq. (7.8.29) and [18]) to prove the identity (1.0.3) $\Gamma(n + 1) = n!$. We give hints of the procedures useful for the following Exercises.

---

[a]The formalism we consider in the following, although easily extendable to the field of complex numbers $\mathbb{C}$, is strictly considered for the real-valued functions (except where it is differently specified).

Proof. Set

$$\Gamma(n+1) = I_n = \int_0^\infty e^{-t}t^n dt, \qquad (7.1.1)$$

and proceed as follows[b]:

$$\sum_{n=0}^\infty \frac{\xi^n}{n!}I_n = \int_0^\infty e^{-t(1-\xi)} dt = \frac{1}{1-\xi}. \qquad (7.1.2)$$

The r.h.s. is expanded in series, thus finding

$$\sum_{n=0}^\infty \frac{\xi^n}{n!}I_n = \sum_{m=0}^\infty \xi^m. \qquad (7.1.3)$$

The comparison between power-like terms eventually yields the proof of the exercise.

$$\frac{\Gamma(n+1)}{n!} = 1. \qquad (7.1.4)$$

$\square$

2 Prove that

$$\int_0^\infty e^{-t^m} dt = \frac{1}{m}\Gamma\left(\frac{1}{m}\right), \qquad \text{Re}(m) > 0. \qquad (7.1.5)$$

Hint: Set $t^m = \sigma$ and note that Eq. (7.1.5) becomes $\frac{1}{m}\int_0^\infty e^{-\sigma}\sigma^{\frac{1}{m}-1}d\sigma$.

3 Use the integration by parts to prove the well-known property

$$\Gamma(n+1) = \int_0^\infty e^t t^n dt = n\Gamma(n). \qquad (7.1.6)$$

4 Use the definition of Γ-function to prove the following Laplace transform identity (see its use, e.g., in Eq. (3.1.2)):

$$A^{-\nu} = \frac{1}{\Gamma(\nu)}\int_0^\infty e^{-sA}s^{\nu-1}ds, \qquad \text{Re}(A), \text{Re}(\nu) > 0. \qquad (7.1.7)$$

---

[b]The summation and integral can be interchanged because the series $\sum_{n=0}^\infty \frac{(\xi t)^n}{n!}$ converges uniformly to $e^{t\xi}$ in the whole integration interval.

5   Prove the identity

$$\int_0^x dx_1 \int_0^{x_1} dx_2 \cdots \int_0^{x_{m-1}} x_m^\mu \, dx_m = \frac{\Gamma(\mu+1)}{\Gamma(m+\mu+1)} x^{m+\mu}.$$
(7.1.8)

6   Use the iterated integration by parts to show that

$$\int_0^x e^{-t} t^{\nu-1} dt = \sum_{s=0}^\infty \frac{\Gamma(\nu)}{\Gamma(\nu+s+1)} x^{\nu+s} e^{-x}.$$
(7.1.9)

Hint: If you do not succeed, do not despair, and look at Section 2.5!

7   Use the same procedure to show that

$$\int_0^1 (1-z)^n z^{\nu-1} dz = \sum_{s=0}^n \frac{\Gamma(\nu)}{\Gamma(s+\nu+1)} \frac{n!}{(n-s)!} z^{s+\nu} (1-z)^{n-s} \Big|_0^1$$

$$= n! \frac{\Gamma(\nu)}{\Gamma(n+\nu+1)} = B(n+1,\nu).$$
(7.1.10)

8   Prove that

$$\frac{\Gamma(\nu)}{\Gamma(\nu+n+1)} = \frac{1}{\nu(\nu+1)\ldots(\nu+n)}.$$
(7.1.11)

9   Prove the equivalence between (1.0.1) and (1.0.2) for $\mathrm{Re}(\nu) > 0$.

Hint: Replace the exponential in the integrand of (1.0.1) by

$$e^{-t} = \lim_{n\to\infty} \left(1 - \frac{t}{n}\right)^n,$$
(7.1.12)

write

$$\Gamma(\nu) = \int_0^\infty e^{-t} t^{\nu-1} dt = \lim_{n\to\infty} n^\nu \int_0^1 (1-z)^n z^{\nu-1} dz. \quad (7.1.13)$$

Use Eqs. (7.1.10)–(7.1.11) to obtain

$$\Gamma(\nu) = \lim_{n\to\infty} n^\nu \int_0^1 (1-z)^n z^{\nu-1} dz = \lim_{n\to\infty} \frac{n^\nu n!}{\nu(\nu+1)\ldots(\nu+n)}.$$
(7.1.14)

10 Find an expression in terms of Γ-function, for the integral[c]

$$I_{\ln} = \int_0^\infty e^{-t} t^{\nu-1} \ln(t)\, dt. \qquad (7.1.15)$$

Hint: Set

$$\ln(t) = \lim_{\delta \to 0} \frac{t^\delta - 1}{\delta} \qquad (7.1.16)$$

and get (use convergence argument of the integrand to justify the interchange of integral and limit operations)

$$I_{\ln} = \lim_{\delta \to 0} \frac{1}{\delta} \int_0^\infty e^{-t} t^{\nu-1} \left(t^\delta - 1 \right) dt = \lim_{\delta \to 0} \frac{\Gamma(\nu + \delta) - \Gamma(\nu)}{\delta}$$
$$= \Gamma'(\nu)$$

$$(7.1.17)$$

where the apex denotes derivative[d] with respect to ν.

11 Use the same argument to prove that

$$I_{l^2} = \int_0^\infty e^{-t} t^{\nu-1} \left(\ln(t) \right)^2 dt = \Gamma''(\nu) \qquad (7.1.18)$$

and

$$I_{l^2} = \int_0^\infty e^{-t^2} \ln(t)\, dt = \frac{1}{4}\Gamma'\left(\frac{1}{2}\right). \qquad (7.1.19)$$

Hint: Employ a different procedure instead of Eq. (7.1.16). Set $t^2 = \sigma$ and write

$$I_{l^2} = \frac{1}{4} \int_0^\infty e^{-\sigma} \sigma^{\frac{1}{2}-1} \ln(\sigma)\, dt \qquad (7.1.20)$$

and then apply Eq. (7.1.17).

---

[c]We denote with "log" the logarithm in base 10 and with "ln" the corresponding case in base e.
[d]We recall that $\Gamma'(\nu) = \psi(\nu)\Gamma(\nu)$ with $\psi(\nu)$ the digamma function or polygamma of order 0.

12 Use the argument outlined before to prove the identities

$$\int_0^\infty e^{-t}\ln(t)dt = \lim_{\delta\to 0,n\to\infty} n\int_0^1 (1-z)^n \frac{(nz)^\delta - 1}{\delta}dz$$

$$= \lim_{\delta\to 0,n\to\infty} \sum_{s=0}^n (-1)^s \binom{n}{s}\frac{n}{\delta}\left(\frac{n^\delta}{s+\delta+1} - \frac{1}{s+1}\right)$$

$$= -\lim_{n\to\infty} \sum_{s=0}^n (-1)^s \binom{n}{s}n\left(\frac{(s+1)\ln(n)-1}{(s+1)^2}\right).$$

$$(7.1.21)$$

13 Show that the limit, for large n of the last of Eq. (7.1.21), is the Euler–Mascheroni constant γ, namely

$$\lim_{n\to\infty} \sum_{s=0}^n (-1)^s \binom{n}{s}n\left(\frac{(s+1)\ln(n)-1}{(s+1)^2}\right)$$

$$= \lim_{n\to\infty} \sum_{k=1}^n \frac{1}{k} - \ln(n) = \gamma \simeq 0.577216\ldots \quad (7.1.22)$$

14 Use Eq. (7.1.14) to prove that

$$\Gamma(\nu) = \frac{1}{\nu}e^{-\gamma\nu}\frac{\prod_{n=1}^\infty e^{\frac{\nu}{n}}}{\prod_{n=1}^\infty \left(1+\frac{\nu}{n}\right)}. \qquad (7.1.23)$$

Hint: Use the following chain of identities:

$$\lim_{n\to\infty} \frac{n^\nu n!}{\nu(\nu+1)\ldots(\nu+n)} = \frac{1}{\nu}\lim_{n\to\infty} e^{\nu\ln(n)}\frac{1}{\prod_{s=1}^n \left(1+\frac{\nu}{s}\right)}$$

$$= \frac{1}{\nu}\lim_{n\to\infty} \exp\left\{\nu\left(-\gamma + \lim_{k\to\infty}\sum_{k=1}^n \frac{1}{k}\right)\right\}\frac{1}{\prod_{s=1}^n \left(1+\frac{\nu}{s}\right)}$$

$$= \frac{1}{\nu}\lim_{n\to\infty} e^{-\gamma\nu}\frac{\prod_{k=1}^n e^{\frac{\nu}{k}}}{\prod_{s=1}^n \left(1+\frac{\nu}{s}\right)} = \frac{1}{\nu}e^{-\gamma\nu}\frac{\prod_{n=1}^\infty e^{\frac{\nu}{n}}}{\prod_{n=1}^\infty \left(1+\frac{\nu}{n}\right)}.$$

$$(7.1.24)$$

15 Use Eq. (7.1.23) to get

$$\frac{1}{\Gamma(x)\Gamma(-x)} = -x^2 \prod_{n=1}^{\infty} \left(1 - \left(\frac{x}{n}\right)^2\right). \qquad (7.1.25)$$

16 Prove that

$$\prod_{n=1}^{\infty} \left(1 - \left(\frac{x}{n}\right)^2\right) = \frac{\sin(\pi x)}{\pi x}. \qquad (7.1.26)$$

Hint: A function $f(x)$ having n real zeros $x_1, 1_2, \ldots, x_n$ can always be constructed as

$$f(x) = \prod_{s=1}^{n}(x - x_s). \qquad (7.1.27)$$

The statement holds for infinite zeros and regarding the function on the r.h.s. of Eq. (7.1.26). It has infinite zeros located at $x = \pm 1, \pm 2, \ldots, \pm n$, it can accordingly be written as

$$\prod_{n=1}^{\infty} \left(1 - \frac{x}{n}\right)\left(1 + \frac{x}{n}\right). \qquad (7.1.28)$$

17 Establish the reflection property

$$\Gamma(x)\Gamma(1 - x) = \frac{\pi}{\sin(\pi x)}. \qquad (7.1.29)$$

Hint: Use the identity $\Gamma(-x) = -\frac{\Gamma(1-x)}{x}\ldots$

18 Prove the identity

$$\Gamma(x) = 2 \int_{0}^{\infty} e^{-u^2} u^{2x-1} du. \qquad (7.1.30)$$

Hint: Set $u^2 = t\ldots$

19 Use the previous identity to state that

$$\int_{0}^{\frac{\pi}{2}} \cos(\theta)^{2x-1} \sin(\theta)^{2y-1} d\theta = \frac{1}{2}B(x, y). \qquad (7.1.31)$$

Hint: Establish first that

$$\Gamma(x)\Gamma(y) = 4 \int_0^\infty \int_0^\infty e^{-(u^2+v^2)} u^{2x-1} v^{2y-1} du\, dv, \quad (7.1.32)$$

then transform to polar coordinates

$$u = \rho\cos(\theta), \qquad v = \rho\sin(\theta) \qquad (7.1.33)$$

and find (pay attention to the upper integration limits)

$$\Gamma(x)\bar{}(y) = 4 \int_0^\infty e^{-\rho^2} \rho^{2(x+y)-2} \rho\,d\rho \int_0^{\frac{\pi}{2}} \cos(\theta)^{2x-1} \sin(\theta)^{2y-1} d\theta$$

$$= 2\Gamma(x+y) \int_0^{\frac{\pi}{2}} \cos(\theta)^{2x-1} \sin(\theta)^{2y-1} d\theta \dots (7.1.34)$$

20 Prove the following identity:

$$\int_C^\infty \frac{x^{\nu-1}}{1+x^2}\, dx = \frac{\pi}{2} \csc\left(\pi\, \frac{\nu}{2}\right), \qquad 0 < \nu < 2. \qquad (7.1.35)$$

Hint: Use the Laplace transform and write

$$\int_0^\infty \frac{x^{\nu-1}}{1+x^2}\, dx = \int_0^\infty e^{-s} \int_0^\infty x^{\nu-1} e^{-sx^2} dx\, ds$$

$$= \frac{1}{2}\Gamma\left(\frac{\nu}{2}\right) \int_0^\infty e^{-s} s^{-\frac{\nu}{2}} ds = \frac{1}{2}\Gamma\left(\frac{\nu}{2}\right)\Gamma\left(1 - \frac{\nu}{2}\right).$$

$$(7.1.36)$$

Apply the reflection property (7.1.29) to get the result.

21 Use the same procedure to get, $\forall \nu, \alpha, \beta \in \mathbb{R} : \alpha > 0$ and $0 < \nu < \alpha\beta$,

$$\int_0^\infty \frac{x^{\nu-1}}{(1+x^\alpha)^\beta}\, dx = \frac{\Gamma\left(\frac{\nu}{\alpha}\right)\Gamma\left(\beta - \frac{\nu}{\alpha}\right)}{\alpha\Gamma(\beta)}. \qquad (7.1.37)$$

Hint: $\int_0^\infty \frac{x^{\nu-1}}{(1+x^\alpha)^\beta} dx = \frac{1}{\Gamma(\beta)} \int_0^\infty e^{-s} s^{\beta-1} \int_0^\infty x^{\nu-1} e^{-sx^\alpha} dx\, ds \dots$

22 Prove the identity

$$\int_0^\infty \frac{t^{x-1}}{(1+t)^{x+y}} dt = B(x, y), \qquad x, y > 0. \qquad (7.1.38)$$

Hint: Apply the following steps:

$$\int_0^\infty \frac{t^{x-1}}{(1+t)^{x+y}}dt = \frac{1}{\Gamma(x+y)}\int_0^\infty e^{-s}s^{x+y-1}\int_0^\infty e^{-st}t^{x-1}dtds$$

$$= \frac{\Gamma(x)}{\Gamma(x+y)}\int_0^\infty e^{-s}s^{y-1}ds\ldots \qquad (7.1.39)$$

23 Use the integral representation in Eq. (7.1.38) to prove the equivalent form

$$\int_0^1 u^{x-1}(1-u)^{y-1}du = B(x,y). \qquad (7.1.40)$$

Hint: Set $t = \frac{u}{1-u}$ in the integral on the r.h.s. of Eq. (7.1.38) and find $\int_0^\infty \frac{t^{x-1}}{(1+t)^{x+y}}dt = \int_0^1 \frac{u^{x-1}}{(1-u)^{x-1}}\frac{(1-u)^{x+y}}{(1-u)^2}du\ldots$

Albeit not strictly related to this complementary part devoted to Euler Beta and Gamma functions, we like to propose the following exercise.

24 Use the notions outlined so far, to prove the following identity[e]:

$$\int_{-\infty}^\infty \frac{\cos(x)}{1+x^2}dx = \frac{\pi}{e}. \qquad (7.1.41)$$

Hint: Before proceeding we mention the identity (see, e.g., [3], Ch. 3, p. 100 and Ref. [18], Ch. 2, pp. 67–68)

$$\int_0^\infty e^{-a^2x^2-\frac{b^2}{x^2}}dx = \frac{1}{2}\frac{\sqrt{\pi}}{a}e^{-2ab}, \qquad (7.1.42)$$

---

[e]Referring to it, Paul J. Nahin, in his beautiful book *An Imaginary Tale. The Story of $\sqrt{-1}$* (Princeton University Press, 1998), noted *"Such calculations were to me, then, seemingly possible only if one had the power of a sorcerer"*. The relevant computation is usually accomplished using the methods of complex analysis, developed by Louis Agustin Cauchy in the 19th century. You are in position to further impress professor Nahin by proving it without any recourse to the technicalities of Complex Analysis.

after which the proof (7.1.41) proceeds as follows:

$$\int_{-\infty}^{\infty} \frac{\cos(x)}{1+x^2}\,dx = \mathrm{Re}\left(\int_0^{\infty} e^{-s}\left(\int_{-\infty}^{\infty} e^{ix-sx^2}\,dx\right)ds\right)$$

$$= \sqrt{\pi}\int_0^{\infty}\frac{e^{-s-\frac{1}{4s}}}{\sqrt{s}}\,ds$$

$$= 2\sqrt{\pi}\int_0^{\infty} e^{-\sigma^2-\frac{1}{4\sigma^2}}\,d\sigma = \frac{\pi}{e}. \qquad (7.1.43)$$

25 Use the previous result to establish the Fourier tranform of a *Lorentzian function* [94].

Hint: Write the Lorentzian as

$$L(x,\alpha) = \frac{2\alpha}{x^2+\alpha^2} \qquad (7.1.44)$$

then proceed as follows:

$$\tilde{L}(k,\alpha) = \frac{2\alpha}{\sqrt{2\pi}}\int_{-\infty}^{\infty}\frac{e^{-ikx}}{\alpha^2+x^2}\,dx$$

$$= \frac{2\alpha}{\sqrt{2\pi}}\int_{-\infty}^{\infty} e^{-ikx}\left(\int_0^{\infty} e^{-sx^2-s\alpha^2}\,ds\right)dx$$

$$= \sqrt{2}\alpha\int_0^{\infty}\frac{e^{-s\alpha^2-\frac{k^2}{4s}}}{\sqrt{s}}\,ds$$

$$= 2\sqrt{2}\alpha\int_0^{\infty} e^{-\alpha^2\sigma^2-\frac{k^2}{4\sigma^2}}\,d\sigma = \sqrt{2\pi}e^{-|k|\alpha}. \quad (7.1.45)$$

26 Prove that, $\forall a,b \in \mathbb{R} : a > 0$

$$I_{cL}(a,b) = \int_{-\infty}^{\infty}\frac{\cos(bx)}{1+ax^2}\,dx = \frac{\pi}{\sqrt{a}}e^{-\frac{b}{\sqrt{a}}}. \qquad (7.1.46)$$

and

$$I_{cL2}(a) = \int_{-\infty}^{\infty}\frac{x^2\cos(x)}{(1+ax^2)^2}\,dx = \frac{\pi}{2a^2}\left(\sqrt{a}-1\right)e^{-\frac{1}{\sqrt{a}}}. \quad (7.1.47)$$

Hint: The first is a straightforward consequence of the previous identities, while the second is obtained from the first after noting that $\frac{d}{da}I_{cL}(a,1) = \dots$

27 Use the outlined procedure, or whatever, to prove the identity (see also [58])

$$I_{eL}(a, b) = \int_{-\infty}^{\infty} \frac{e^{-ax^2}}{1 + bx^2} dx = \sqrt{\frac{\pi}{b}} e^{\frac{a}{b}} \Gamma\left(\frac{1}{2}, \frac{a}{b}\right), \quad (7.1.48)$$

where

$$\Gamma(\nu, x) = \int_{x}^{\infty} e^{-t} t^{\nu-1} dt = \Gamma(\nu) - \gamma(\nu, x), \quad (7.1.49)$$

is the *upper incomplete gamma function* and

$$\gamma(\nu, x) = \int_{0}^{x} e^{-t} t^{\nu-1} dt \quad (7.1.50)$$

the *lower incomplete gamma function*.

Hint: Apply the steps reported as follows.

$$I_{eL}(a, b) = \int_{-\infty}^{\infty} \frac{e^{-ax^2}}{1 + bx^2} dx = \int_{0}^{\infty} e^{-s} \left(\int_{-\infty}^{\infty} e^{-(a+sb)x^2} dx \right) ds$$

$$= \sqrt{\pi} \int_{0}^{\infty} \frac{e^{-s}}{\sqrt{a + sb}} ds = \frac{\sqrt{\pi}}{b} e^{\frac{a}{b}} \int_{a}^{\infty} \frac{e^{-\frac{\tau}{b}}}{\sqrt{\tau}} d\tau. \quad (7.1.51)$$

28 Consider the function (an extension of the Γ-function)

$$\Gamma_2(x, \nu) = \int_{0}^{\infty} e^{-t^2} e^{-xt} t^{\nu-1} dt, \quad 0 < \nu < 1, \quad (7.1.52)$$

show that it can be expanded in series as

$$\Gamma_2(x, \nu) = \frac{1}{2} \sum_{r=0}^{\infty} \Gamma\left(\frac{r+\nu}{2}\right) \frac{(-x)^r}{r!} \quad (7.1.53)$$

and study the relevant convergence conditions.

Hint: Note

$$\Gamma_2(x, \nu) = \int_{0}^{\infty} e^{-t^2} t^{\nu-1+x\partial_x} dt \, e^{-x} = \Gamma\left(\frac{\nu + x\partial_x}{2}\right) e^{-x} \dots$$

then use identity (7.3.4).

29 Use the Borel operator (1.1.19) to show that[f]

$$\int_0^\infty e^{-t} H_n(xt)dt = n! \sum_{r=0}^{\lfloor \frac{n}{2} \rfloor} \frac{(-1)^r (2x)^{n-2r}}{r!}. \qquad (7.1.54)$$

Hint: According to the later Section 7.3, Eq. (7.3.4), we write

$$\Gamma(x\partial_x + 1)H_n(x) = n! \sum_{r=0}^{\lfloor \frac{n}{2} \rfloor} \Gamma(n - 2r + 1)\frac{(-1)^r (2x)^{n-2r}}{(n - 2r)!r!}.$$
$$(7.1.55)$$

30 Prove the following identity concerning the two-variable Hermite polynomials:

$$\Pi_n(x, y) = \int_0^\infty e^{-t} H_n(xt, yt)dt = \Gamma(x\partial_x + y\partial_y + 1)H_n(x, y)$$

$$= n! \sum_{r=0}^{\lfloor \frac{n}{2} \rfloor} \frac{(n - r)! \, x^{n-2r}y^r}{(n - 2r)!r!}. \qquad (7.1.56)$$

31 Show that the polynomials $\Pi_n(x, y)$, for $\mathrm{Re}(\tau(x - y\tau)) < 1$, satisfy the generating function

$$\sum_{n=0}^{\infty} \frac{\tau^n}{n!} \Pi_n(x, y) = \int_0^\infty e^{-(1 - x\tau - y\tau^2)t} dt = \frac{1}{1 - x\tau - y\tau^2}.$$
$$(7.1.57)$$

The elements of discussion we have developed so far have been aimed at proving how helpful the use of the Euler functions can be to solve a variety of problems involving integrals, wildly appearing in almost any field of Applied Mathematics.

7.2   Deeper Inside Gaussians

We have underscored that the use of umbral methods is a "tool" allowing the "downgrade" of the degree of complexity of a special function. The Gaussian can indeed be formally written as a rational

---

[f]The one-variable Hermite polynomials $H_n(x)$ will be specifically treated in the following sections.

function, the Bessel as a Gaussian and so on. We have also noted that the roots of this point of view can be traced back to the conceptual context of Borel transform. We use here the rational umbral image of Gaussian to get a deeper understanding of the relevant role in the realm of special functions.

32 Prove the downgrading of the Gaussian to a rational function[g]

$$e^{-x^2} = \frac{1}{1 + \hat{c}x^2} \varphi_0 \tag{7.2.1}$$

where $\hat{c}$ is the operator introduced in (1.1.6).

Hint: The proof is fairly straightforward and follows by a naïve series expansion of the r.h.s. .

33 Use Eq. (7.2.1) to prove that[h]

$$\int e^{-s^2} ds = \sqrt{\hat{c}} \arctan\left(\sqrt{\hat{c}} s\right) \varphi_0 \tag{7.2.2}$$

and write the result after the operator $\hat{c}$ has acted.

A naïve comment, regarding this result, is that the umbral image of the primitive of a Gaussian is an elementary function, even though this is not true for the non umbral counterpart. This is just a consequence of the function transcendence downgrading.

Definition 34. We introduce the function[i]

$$C_g(x) = \frac{1}{2}\left(\frac{1}{1 - i\hat{c}^{\frac{1}{2}}x} + \frac{1}{1 + i\hat{c}^{\frac{1}{2}}x}\right)\varphi_0, \qquad \forall x \in \mathbb{R} \tag{7.2.3}$$

and the associated function

$$S_g(x) := \frac{1}{2i}\left[\frac{1}{1 - i\hat{c}^{\frac{1}{2}}x} - \frac{1}{1 + i\hat{c}^{\frac{1}{2}}x}\right]\varphi_0, \qquad \forall x \in \mathbb{R} \tag{7.2.4}$$

The above functions (7.2.3)–(7.2.4) will be referred to as cosine and sine Gaussian functions, respectively.

---

[g]We have already exploited this umbral definition for the Gaussian, as, e.g., in Eq. (1.3.9).
[h]From now on, in dealing with undefined integrals, we will omit the constant k in the solution giving it by subunderstanding.
[i]Index g is for Gaussian.

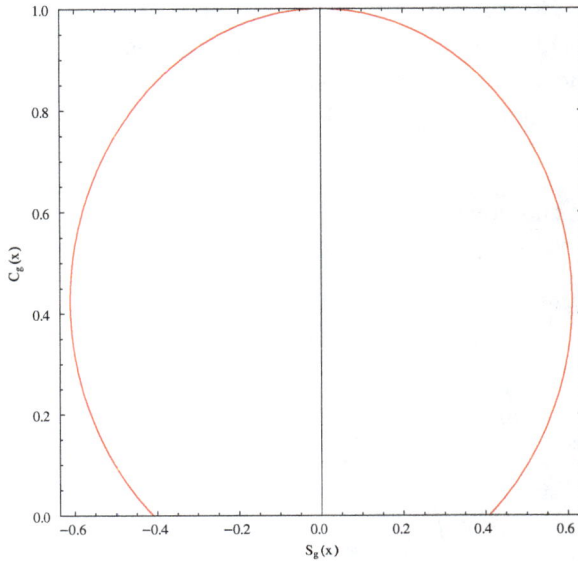

Figure 7.1: Gaussian trigonometric "circumference": egg-shaped curve. $C_g(x)$ vs $S_g(x)$.

In Fig. 7.1, we have reported the Gaussian trigonometric circle along with the geometrical definition of the associated sine and cosine functions and the relevant behaviors vs x, calculated according to Exercises 34 and 35, are reported in Fig. 7.2.

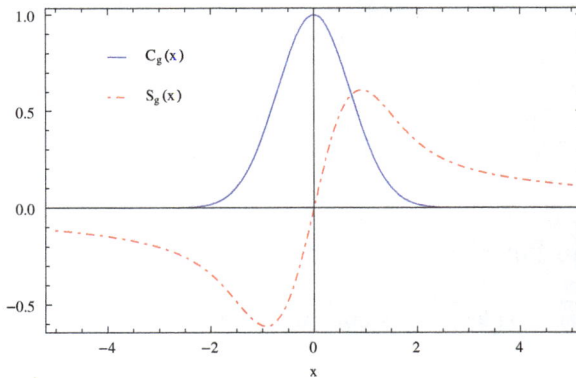

Figure 7.2: $C_g(x)$ and $S_g(x)$ vs x.

34 Prove that the function $C_g(x)$ is equal to the Gaussian e^{-x^2}.

35 Derive the analytical form of the Gaussian sine.

Hint: Use the umbral definition and the series expansion to obtain

$$S_g(x) = \cdots = \sum_{r=0}^{\infty} (-1)^r \frac{x^{2r+1}}{\Gamma\left(r + \frac{3}{2}\right)} = e^{-x^2} \text{erfi}(x), \qquad (7.2.5)$$

where $Erfi(x) = \frac{\text{erf}(ix)}{i}$ and

$$\text{erf}(x) = \frac{2}{\sqrt{\pi}} \int_0^x e^{-t^2} dt \qquad (7.2.6)$$

is the *error funtion*, proportional to the primitive of the Gaussian.

36 Verify that Gaussian sine and cosine satisfy the Kramers–Kronig identity[j] [106]

$$S_g(x) = -\frac{1}{\pi} \mathcal{P} \int_{-\infty}^{\infty} \frac{C_g(\xi)}{\xi - x} d\xi, \qquad (7.2.7)$$

where $\mathcal{P}$ is the Cauchy principal value

$$\mathcal{P} \int_{-\infty}^{\infty} \frac{C_g(\xi)}{\xi - x} d\xi = \lim_{\varepsilon \to 0+} \left[\int_{-\infty}^{x-\varepsilon} \frac{C_g(\xi)}{\xi - x} d\xi + \int_{x+\varepsilon}^{\infty} \frac{C_g(\xi)}{\xi - x} d\xi \right].$$
$$(7.2.8)$$

37 Use the Laplace transform method to derive the following integral representations:

$$C_g(x) = \int_0^{\infty} e^{-s} \cos\left(\sqrt{\hat{c}}\, x\, s\right) ds\, \varphi_0,$$
$$\qquad\qquad (7.2.9)$$
$$S_g(x) = \int_0^{\infty} e^{-s} \sin\left(\sqrt{\hat{c}}\, x\, s\right) ds\varphi_0.$$

---

[j]Identities of the type (7.2.7) are known as Hilbert transform to mathematicians and play a pivotal role in the Theory of Signals or in the study of complex analytical functions.

Hint: Note that the first equation simply follows by the Laplace transform $\frac{k}{k^2+a^2} = \int_0^\infty e^{-kt} \cos(at)dt$ and then:

$$C_g(x) = \frac{1}{1 + (\sqrt{\hat{c}}\, x)^2} \varphi_0 = \int_0^\infty e^{-s} \cos\left(\sqrt{\hat{c}}\, x\, s\right) ds\, \varphi_0.$$

$$(7.2.10)$$

38 Show that the successive derivatives of the Gaussian trigonometric functions can be written as

$$C_g^{(m)}(x) = \sqrt{\hat{c}^m} \int_0^\infty e^{-s} s^m \cos\left(\sqrt{\hat{c}}\, x\, s + m\frac{\pi}{2}\right) ds\, \varphi_0,$$

$$S_g^{(m)}(x) = \sqrt{\hat{c}^m} \int_0^\infty e^{-s} s^m \sin\left(\sqrt{\hat{c}}\, x\, s + m\frac{\pi}{2}\right) ds\, \varphi_0.$$

$$(7.2.11)$$

Hint: Recall that $\left(\frac{d}{dx}\right)^m \cos(x) = \cos\left(x + m\frac{\pi}{2}\right)$.

39 Find a link between Hermite polynomials $H_m(x)$ and the mth derivative of the Gaussian cosine $C_g^{(m)}(x)$.

Hint: Recall that

$$\left(\frac{d}{dx}\right)^m e^{-x^2} = (-1)^m H_m(x) e^{-x^2}.$$

$$(7.2.12)$$

40 Define the complex function

$$\tilde{E}_g(x) := \frac{1 + i\,\hat{c}^{\frac{1}{2}}\, x}{1 + \hat{c}x^2}\, \varphi_0\,.$$

$$(7.2.13)$$

and show that the Gaussian cosine and sine are its real and imaginary parts.

41 Show that the complex g-function $\tilde{E}_g(x)$ is linked to the Mittag–Leffler[k] polynomial $E_{\alpha,\beta}(x)$ (see Eq. (1.3.1)) in the following way[l]

$$\tilde{E}_g(x) = E_{\frac{1}{2},1}(ix) := E_{\frac{1}{2}}(x).$$

$$(7.2.14)$$

---

[k]We reserved a more substantive section to ML function in the following.
[l]When $\beta = 1$, we use the notation $E_{\alpha,1}(x) := E_\alpha(x)$ for conciseness.

42 It is worth noting that

$$e^{-x^2} = \frac{1}{2}(E_{\frac{1}{2}}(ix) + E_{\frac{1}{2}}(-ix)), \qquad (7.2.15)$$

which appears to be a particular case of a more general identity. We note indeed that the umbral formalism yields Eq. (1.3.2) $E_\alpha(x) = \frac{1}{1-\hat{c}^\alpha x}\varphi_0$ and therefore we easily infer that

$$E_{2\alpha}(x) = \frac{1}{2}\left(E_\alpha(\sqrt{x}) + E_\alpha(-\sqrt{x})\right). \qquad (7.2.16)$$

Work out the explicit proof.

43 Use Eq. (1.3.2) to prove the identity

$$E_\alpha(x) = 1 + xE_{\alpha,\alpha+1}(x). \qquad (7.2.17)$$

Hint: Write $E_\alpha(x) = \frac{(1-\hat{c}^\alpha x)+\hat{c}^\alpha x}{1-\hat{c}^\alpha x}\varphi_0$.

44 Apply the same procedure to state

$$E_\alpha(-x) = E_{2\alpha}(x^2) - xE_{2\alpha,\alpha+1}(x^2). \qquad (7.2.18)$$

Hint: $E_\alpha(-x) = \dfrac{1}{1+\hat{c}^\alpha x} = \dfrac{1-\hat{c}^\alpha x}{1-\hat{c}^{2\alpha} x^2}\varphi_0 = \dots$

45 Apply the techniques exploited so far to justify the identities

$$\frac{1}{(1-\hat{c}^\alpha x)(1+3\,\hat{c}^\alpha x)}\varphi_0 = \frac{1}{4}\left(E_\alpha(x) + 3E_\alpha(-3x)\right) \quad (7.2.19)$$

and

$$\frac{1}{(1-\hat{c}^\alpha x)(1-3\,\hat{c}^\alpha x)}\varphi_0 = \frac{1}{2}\left(-E_\alpha(x) + 3E_\alpha(3x)\right) \quad (7.2.20)$$

Hint: Use the method of partial fraction.

46 Use the umbral form (7.2.1) of the Gaussian function and evaluate the relevant Fourier transform

$$\tilde{f}(k) = \frac{1}{\sqrt{2\pi}}\int_{-\infty}^{\infty}\frac{e^{-ikx}}{1+\hat{c}x^2}dx\,\varphi_0. \qquad (7.2.21)$$

Hint: By using the Laplace transform...

$$\frac{1}{\sqrt{2\pi}} \int_{-\infty}^{\infty} \frac{e^{-ikx}}{1+\hat{c}x^2}dx \,\varphi_0 = \cdots = \frac{1}{\sqrt{2\,\hat{c}}} \int_0^{\infty} \frac{e^{-s-\frac{k^2}{4s\hat{c}}}}{\sqrt{s}}ds \,\varphi_0$$

$$= \sqrt{\frac{2}{\hat{c}}} \int_0^{\infty} e^{-\sigma^2 - \frac{k^2}{4\hat{c}\sigma^2}} d\sigma \,\varphi_0 = \sqrt{\frac{\pi}{2\hat{c}}} e^{-\frac{|k|}{\sqrt{\hat{c}}}} \varphi_0 = \cdots \quad (7.2.22)$$

47 Consider the Bessel–Wright function $W_\beta^\alpha(x) = \sum_{r=0}^{\infty}$ $\frac{x^r}{r!\Gamma(\alpha r - \beta + 1)}$ in Eq. (1.1.1) and evaluate the following "Fourier-like" transform identity

$$\tilde{f}^\alpha(k) = \frac{1}{\sqrt{2\pi}} \int_{\infty}^{\infty} e^{-x^2} W_0^{\frac{\alpha}{2}}(-ikx)dx, \quad (7.2.23)$$

and prove that

$$\tilde{f}^\alpha(k) = \frac{1}{\sqrt{2}} W_0^\alpha \left(-\left(\frac{k}{2}\right)^2 \right). \quad (7.2.24)$$

Hint: Use the umbral representation $W_0^\alpha(x) = e^{\hat{\partial}^\alpha x}\varphi_0 = \cdots$ in Eq. (1.1.3).

48 Use the following notation for a Gaussian integral (see [58]):

$$I_G(a,b) = \int_{-\infty}^{\infty} e^{-(ax^2+bx)}dx, \quad (7.2.25)$$

and prove that

$$I_J(a,b) = \int_{-\infty}^{\infty} J_0\left(2\sqrt{ax^2+bx}\right)dx = I_G(\hat{c}a, \hat{c}b)\varphi_0$$

$$= \sqrt{\frac{\pi}{a}}\sqrt{\frac{b}{2\sqrt{a}}} \, I_{-\frac{1}{2}}\left(\frac{b}{\sqrt{a}}\right), \quad (7.2.26)$$

where $J_0(x)$ is the zeroth-order Bessel function used so much, $\hat{c}$ the umbral operator already used in this chapter and $I_\nu(x)$ the modified Bessel of the first kind in (4.3.1).

Hint: Bessel integral can be written as

$$I_J(a,b) = \int_{-\infty}^{\infty} e^{-((\hat{c}a)x^2+(\hat{c}b)x)}dx \,\varphi_0, \quad (7.2.27)$$

which is sufficient to prove the first part of the exercise. Furthermore, note that

$$I_G(\hat{c}a, \hat{c}b)\varphi_0 = \sqrt{\frac{\pi}{\hat{c}a}} e^{\frac{\hat{c}b^2}{4a}} \varphi_0, \qquad (7.2.28)$$

expand it and use the expansion in series of $I_\nu(x)$ in (4.3.1) (see also Section 2.4 about the application of this procedure).

7.2.1 Hermite Calculus

49 In [31], the following integral:

$$J(a, b, c) = \int_{-\infty}^{\infty} e^{-(ax^4 + bx^2 + cx)} dx, \qquad \forall b, c \in \mathbb{R}, \forall a \in \mathbb{R}_0^+,$$
$$(7.2.29)$$

has been considered, within the framework of problems regarding the non-perturbative treatment of the anharmonic oscillator. A possible perturbative treatment is that of setting

$$J(a, b, c) = \sum_{n=0}^{\infty} \frac{(-1)^n}{n!} g_n(a) \left[H_n(c, -b) + H_n(-c, -b) \right],$$

$$g_n(a) = \int_0^{\infty} x^n e^{-ax^4} dx = \frac{1}{4} a^{-\frac{n+1}{4}} \Gamma\left(\frac{n+1}{4}\right), \quad (7.2.30)$$

which, as noted in [31], is an expansion with zero radius of convergence in spite of the fact that $J(a, b, c)$ is an entire function for any real or complex value of b, c. The use of our point of view allows to write[m]

$$J(a, b, c) = \int_{-\infty}^{\infty} e^{-\hat{h}_{(b,-a)}x^2 - cx} dx \, \eta_0 = \sqrt{\frac{\pi}{\hat{h}}} e^{\frac{c^2}{4\hat{h}}} \eta_0$$

$$= \sqrt{\pi} \sum_{s=0}^{\infty} \frac{1}{s!} \left(\frac{c}{2}\right)^{2s} \hat{h}^{-\left(s+\frac{1}{2}\right)} \eta_0. \qquad (7.2.31)$$

---

[m] We have omitted the subscript $(b, -a)$ in the r.h.s. of Eq. (7.2.31) to avoid a cumbersome notation.

The meaning of the operator $\hat{h}$ raised to a negative exponent is easily understood as

$$\hat{h}^{-\left(s+\frac{1}{2}\right)}\eta_0 = H_{-\left(s+\frac{1}{2}\right)}(b, -a), \qquad (7.2.32)$$

where the negative index Hermite polynomials are expressed in terms of the *parabolic cylinder functions* D_n according to the identity [2]

$$H_{-n}(x, -y) = (2y)^{-\frac{n}{2}} e^{\frac{x^2}{8y}} D_{-n}\left(\frac{x}{\sqrt{2y}}\right). \qquad (7.2.33)$$

The use of Eq. (7.2.33) in Eq. (7.2.31) finally yields the same series expansion obtained in [31]

$$J(a, b, c) = \sqrt{\pi} \sum_{s=0}^{\infty} \frac{1}{s!} \left(\frac{c}{2}\right)^{2s} (2a)^{-\frac{1}{2}\left(s+\frac{1}{2}\right)} e^{\frac{b^2}{8a}} D_{-\left(s+\frac{1}{2}\right)}\left(\frac{b}{\sqrt{2a}}\right), \qquad (7.2.34)$$

which is convergent for any value of b, c and $a > 0$.

50　Regarding the use of non-integer Hermite polynomials, it is evident that the definition adopted in Eq. (2.4.19) can be replaced by the use of the parabolic cylinder function, it is therefore worth noting that the use of the properties of the D_n functions allows the following alternative form for Eq. (2.4.18) (see [176] and Eq. (7.2.33)):

$$I(\gamma, \beta) = \sqrt{\pi} \, (2\,\beta)^{-\frac{1}{4}} e^{\frac{\gamma^2}{8\beta}} D_{-\frac{1}{2}}\left(\frac{\gamma}{\sqrt{2\beta}}\right)$$

$$= \sqrt{\frac{\gamma}{2\sqrt{2\beta}}} \, (2\,\beta)^{-\frac{1}{4}} e^{\frac{\gamma^2}{8\beta}} K_{\frac{1}{4}}\left(\frac{\gamma^2}{8\beta}\right), \qquad (7.2.35)$$

$$D_{-\frac{1}{2}}(z) = \sqrt{\frac{z}{2\pi}} K_{\frac{1}{4}}\left(\frac{1}{4}z^2\right),$$

where $K_\nu(z)$ is a *modified Bessel function of the second kind* (see Chapter 4).

51 Prove the following result:

$$\int_0^\infty e^{-(\beta x^{2n} + \gamma x^n)} dx = \frac{1}{n} \Gamma \left(\frac{1}{n} \right) (2\beta)^{-\frac{1}{2n}} e^{\frac{\gamma^2}{8\beta}} D_{-\frac{1}{n}} \left(\frac{\gamma}{\sqrt{2\beta}} \right).$$
(7.2.36)

The use of this family of polynomials allows to cast the integral in Eq. (7.2.31) in the form

$$J(a, b, c) = \int_{-\infty}^\infty e^{4\hat{h}_{(-c,-a)} x - b x^2} dx \, \epsilon_0 = \sqrt{\frac{\pi}{b}} e^{\frac{\left(4\hat{h}_{(-c,-a)} \right)^2}{4b}} \epsilon_0,$$
(7.2.37)

where

$$\left(4\hat{h}_{(-c,-a)} \right)^n \epsilon_0 := \epsilon_n = H_n^{(4)}(-c, -a) = (-1)^n n! \sum_{r=0}^{\lfloor \frac{n}{4} \rfloor} \frac{c^{n-4r}(-a)^r}{(n-4r)! r!}.$$
(7.2.38)

The series expansion of the r.h.s. of Eq. (7.2.37) in terms of fourth-order Hermite converges in a much more limited range than the series (7.2.34) and has been presented to emphasize the possibilities of the method we have presented so far.

In this section, we have touched upon different and seemingly uncorrelated topics, ranging from Gaussian/Lorentzian to dispersion relations and to Mittag–Leffler functions. In the forthcoming sections we will see how the formalism can be stretched even more to provide further interesting results.

7.3 Generalized Transforms and Operator Methods

We have introduced in Eq. (1.1.24) the Borel operator, expressed as a Gamma function, having a differential operator as argument. In this section, we will see how this definition and the associated formalism can be exploited to make different and substantive progress in the matter we are studying [80]. The following exercises have been explored throughout the book.

Before entering into the main body of the discussion, we recall a few identities specifying the action of an exponential operator on a given function. It is well known that the shift operator [18], used often in the expressions, acts as

$$e^{\lambda \partial_x} f(x) = f(x + \lambda) \tag{7.3.1}$$

and the proof is easily achieved by expanding, according to Taylor, the exponential and getting

$$e^{\lambda \partial_x} f(x) = \sum_{n=0}^{\infty} \frac{\lambda^n}{n!} \partial_x^n f(x). \tag{7.3.2}$$

The r.h.s. of the previous identity represents the series expansion of $f(x + \lambda)$.

The expansion operator $e^{\lambda x \partial_x} f(x) = f(e^{\lambda} x)$ has been proved in (1.1.14) and the following.

52 State that

$$f(x\partial_x)x^n = f(n)x^n, \tag{7.3.3}$$

which implies (1.1.17), and prove the ricorsive identity

$$(x\partial_x)^p x^n = n^p x^n. \tag{7.3.4}$$

Hint: Expanding in series the operator function, we find

$$f(x\partial_x) = \sum_{r=0}^{\infty} \frac{f_r}{r!} (x\partial_x)^r \tag{7.3.5}$$

and therefore $f(x\partial_x)x^n = \sum_{r=0}^{\infty} \frac{f_r}{r!} n^r x^n \ldots$

53 Assume that the function $f(x)$ admits a Fourier transform and that the associated anti-transform exists. Use the previous identities to prove that

$$f(x\partial_x)g(x) = \frac{1}{\sqrt{2\pi}} \int_{-\infty}^{\infty} \tilde{f}(k) e^{ikx\partial_x} dk \; g(x)$$

$$= \frac{1}{\sqrt{2\pi}} \int_{-\infty}^{\infty} \tilde{f}(k) g\left(e^{ik} x\right) dk. \tag{7.3.6}$$

54 Show that, as to the 0th cylindrical Bessel functions, the shift opertor methods yield (1.2.17) as

$$J_0(x) = e^{-\left(\frac{x}{2}\right)^2 e^{\partial_z}} \frac{1}{\Gamma(z+1)} \Big|_{z=0}$$

$$= \sum_{r=0}^{\infty} \frac{(-1)^r}{r!} \left(\frac{x}{2}\right)^{2r} e^{r\partial_z} \frac{1}{\Gamma(z+1)} \Big|_{z=0}$$

$$= \sum_{r=0}^{\infty} \frac{(-1)^r}{r!} \left(\frac{x}{2}\right)^{2r} \frac{1}{\Gamma(z+r+1)} \Big|_{z=0} = \sum_{r=0}^{\infty} \frac{(-1)^r}{(r!)^2} \left(\frac{x}{2}\right)^{2r}.$$

$$(7.3.7)$$

55 Regarding the integration procedure, we have also noted that (1.2.21)

$$\int_{-\infty}^{\infty} J_0(x)dx = \sqrt{\pi}\, e^{-\frac{1}{2}\partial_z} \frac{1}{\Gamma(z+1)} \Big|_{z=0} = \sqrt{\pi} \frac{1}{\Gamma\left(z+\frac{1}{2}\right)} \Big|_{z=0}$$

$$= \sqrt{\pi} \frac{1}{\left(\frac{\sqrt{\pi}}{2}\right)} = 2. \qquad (7.3.8)$$

56 An interesting byproduct of this point of view is the solution of the *Lamb–Bateman* equation [125] originally proposed to treat the diffraction of a solitary wave [116], namely

$$\int_0^{\infty} u(x - y^2)dy = f(x). \qquad (7.3.9)$$

The unknown function $u(x)$ can be obtained by noting that

$$u(x - y^2) = e^{-y^2 \partial_x} u(x), \qquad (7.3.10)$$

accordingly, we find

$$\left(\int_0^{\infty} e^{-y^2 \partial_x} dy\right) u(x) = f(x). \qquad (7.3.11)$$

Furthermore, since by using (1.2.2) we get

$$\int_0^{\infty} e^{-y^2 \partial_x} dy = \frac{1}{2}\sqrt{\pi}\partial_x^{-\frac{1}{2}}, \qquad (7.3.12)$$

we end up with the solution

$$u(x) = \frac{2}{\sqrt{\pi}} \left(\partial_x^{\frac{1}{2}} f(x) \right),$$ (7.3.13)

namely, the solution of the Lamb–Bateman problem is just the derivative of interger order half- of the function $f(x)$.

57 Use the shift operator method to prove the following identity:

$$\int_{-\infty}^{\infty} e^{-ax^2} f(x + \beta) dx = \sqrt{\frac{\pi}{a}} e^{\frac{1}{4a} \partial_\beta^2} f(\beta).$$ (7.3.14)

Hint: Recall that the operator containing the derivative ∂_β can be treated as an ordinary algebraic quantity.

58 Use the properties of Hermite polynomials and the identity (7.3.14) to state that

$$\int_{-\infty}^{\infty} e^{-ax^2} H_n(x + \beta) dx = \sqrt{\frac{\pi}{a}} H_n \left(2\beta, \frac{1 - \alpha}{\alpha} \right).$$ (7.3.15)

Hint: By using identity *iii)* $H_n(\beta) = H_n(2\beta, -1)$ of Eq. (1.2.9), we find that the integral is equal to $\sqrt{\frac{\pi}{a}} e^{\frac{1}{4a} \partial_\beta^2} H_n$ $(2\beta, -1)$ and, after applying the rule $e^{z \partial_x} H_n(x, y) = H_n(x, y + z)$ [18], the solution follows naïvely.

59 Use the same argument of the previous exercise to prove that

$$\int_{-\infty}^{\infty} e^{-ay^2} H_n(x, y + \beta) dy = \sqrt{\frac{\pi}{a}} e^{\frac{1}{4a} \partial_\beta^2} H_n(x, \beta)$$

$$= \sqrt{\frac{\pi}{a}} n! \sum_{r=0}^{\lfloor \frac{n}{2} \rfloor} \frac{x^{n-2r}}{(n - 2r)! r!} H_r \left(\beta, \frac{1}{4a} \right).$$

(7.3.16)

Hint: The proof of the second part of the exercise follows from

$$e^{\frac{1}{4a} \partial_\beta^2} H_n(x, \beta) = n! \sum_{r=0}^{\lfloor \frac{n}{2} \rfloor} \frac{x^{n-2r}}{(n - 2r)! r!} \left(e^{\frac{1}{4a} \partial_\beta^2} \beta^r \right) \dots$$

60 We invite the reader to consider Exercise 59 from the operatorial point of view, reported as follows:

$$e^{\frac{1}{4a}\partial_\beta^2} H_n(x,\beta) = e^{\frac{1}{4a}\partial_\beta^2} e^{\beta\partial_x^2} x^n = \ldots \qquad (7.3.17)$$

and to obtain a meaning for the r.h.s. of the previous identity.

The following section covers some topics touched upon in the main bodies of the other chapters without the necessary details and extends the *Hermite and Laguerre Calculus.*

7.4 Umbra and Higher-Order Hermite Polynomials

The higher-order Hermite polynomials, also called *Lacunary HP* [18, 72], defined through the operational identity

$$H_n^{(m)}(x,y) = e^{y\partial_x^m} x^n, \qquad \forall m \in \mathbb{N}, \qquad (7.4.1)$$

specified by the series

$$H_n^{(m)}(x,y) = n! \sum_{r=0}^{\lfloor \frac{n}{m} \rfloor} \frac{x^{n-mr} y^r}{(n-mr)! r!}, \qquad (7.4.2)$$

with generating function

$$\sum_{n=0}^{\infty} \frac{t^n}{n!} H_n^{(m)}(x,y) = e^{xt+yt^m}, \qquad (7.4.3)$$

can be reduced to the nth power of a binomial as shown in the following exercise.

61 We introduce the umbral operator ${}_y\hat{h}_m^r$ such that, $\forall r \in \mathbb{R}$,

$$
{}_y\hat{h}_m^r\, {}_m\theta_0 := {}_m\theta_r = \frac{y^{\frac{r}{m}} r!}{\Gamma\left(\frac{r}{m}+1\right)} A_{m,r},
$$

$$
A_{m,r} = \begin{cases} 1 & s \in \mathbb{Z}, \\ & r = ms \\ 0 & s \notin \mathbb{Z}, \end{cases} \qquad (7.4.4)
$$

which allows to define them as

$$H_n^{(m)}(x,y) = \left(x + {}_y\hat{h}_m\right)^n {}_m\theta_0. \qquad (7.4.5)$$

For $m = 2$, we recover Eq. (2.3.3).

62 It is clearly evident that not too much effort is necessary to study the relevant properties and associated functions, which can be derived using the same procedure adopted for the second-order case $(m = 2)$ as, e.g., *higher negative order*

$$H_{-\nu}^{(m)}(x,y) = (x + {}_{-|y|}\hat{h}_m)^{-\nu} {}_m\theta_0$$

$$= \frac{1}{\Gamma(\nu)} \int_0^\infty s^{\nu-1} e^{-sx} e^{-|y|\hat{h}_m s} {}_m\theta_0 \, ds, \, (7.4.6)$$

or

$$H_{-\nu}^{(m)}(x,y) = \frac{1}{\Gamma(\nu)} \int_0^\infty s^{\nu-1} e^{-sx} e^{-ys^m} ds, \quad \text{Re}(y) > 0. \tag{7.4.7}$$

It is therefore evident that, $\forall m, n \in \mathbb{N} : m > n$,

$$I^{(m,1)}(x,y) = \int_0^\infty e^{-sx} e^{-s^m y} ds = H_{-1}^{(m-1)}(x,y),$$

$$I^{(m,2)}(x,y) = \int_0^\infty e^{-s^2 x} e^{-s^m y} ds = \sqrt{\pi}\, H_{-1/2}^{(m-2)}(x,y),$$

$$...$$

$$I^{(m,n)}(x,y) = \int_0^\infty e^{-s^n x} e^{-s^m y} ds = \frac{\Gamma\left(\frac{1}{n}\right)}{n} H_{-1/n}^{(m-n)}(x,y). \tag{7.4.8}$$

63 *Super-Gaussian* (SG) functions are used in Optics to describe the so-called flattened beams. Limiting ourselves to the one-dimensional case, they are represented by an exponential function of the type

$$S_G(x,m) = e^{-x^m}, \qquad (7.4.9)$$

where, for simplicity, m is assumed to be an even integer. The relevant propagation has been treated by the use of effective methods superposition employing Gaussian beams

[98], which amount to the approximation of an SG beam with a superposition of Gauss beam, whose transformation through a lens-like device is well known. The use of an alternative approach (even though less efficient than the flattened beam method) is suggested by the means of a Fresnel transform [33], which reads

$$S_G(x, m; A, B, D) = \frac{1}{\sqrt{2\pi i B}} \int_{-\infty}^{+\infty} e^{\frac{i}{2B}(Ay^2 + 2xy + Dx^2)} S_G(y, m) dy,$$

(7.4.10)

where A, B, D are constants accounting for the optical elements constituting the transport line. It is evident, according to the previous formalism, that the above integral can be cast in the form

$$S_G(x, m; A, B, D)$$
$$= \frac{e^{\frac{iD}{2B}x^2}}{\sqrt{2\pi i B}} \left[H_{-1}^{(4,2)} \left(\frac{ix}{B}, \frac{iA}{2B}, 1 \right) + H_{-1}^{(4,2)} \left(-\frac{ix}{B}, \frac{iA}{2B}, 1 \right) \right],$$

(7.4.11)

where the Hermite function on the right corresponds to the polynomial

$$H_n^{(4,2)}(x, y, z) = n! \sum_{r=0}^{\lfloor \frac{n}{4} \rfloor} \frac{H_{n-4r}(x, y) z^r}{(n-4r)! r!}.$$

(7.4.12)

They are also defined by the operational identity

$$H_n^{(4,2)}(x, y, z) = e^{z \partial_x^4} H_n(x, y),$$

(7.4.13)

and extended to the associated functions defined as

$$H_{-\nu}^{(4,2)}(x, y, z) = \frac{1}{\Gamma(\nu)} \int_0^\infty s^{\nu-1} e^{xt+yt^2-zt^4} ds, \quad \mathrm{Re}(y) > 0.$$

(7.4.14)

The possibility we have envisaged of using Hermite functions to study the propagation of SG beams needs various refinements to become an effective tool, notwithstanding that it provides a further proof of the possibility offered by these techniques.

64 In order to complete the discussion on SG modes, we like to provide the following examples regarding the properties of the Gauss–Weierstrass transform (1.2.3);

$$e^{y\,\partial_x^2} f(x) = \frac{1}{2\sqrt{\pi y}} \int_{-\infty}^{\infty} e^{-\frac{(x-\xi)^2}{4y}} f(\xi)\, d\xi$$

$$= \frac{1}{2\sqrt{\pi y}} \int_{-\infty}^{\infty} e^{-\frac{\xi^2}{4y}} e^{-\frac{x^2}{4y}+\frac{x\xi}{2y}} f(\xi)\, d\xi. \quad (7.4.15)$$

The use of the generating function of Hermite polynomials (1.2.6) allows to cast Eq. (7.4.15) in the form

$$e^{y\,\partial_x^2} f(x) = \frac{1}{2\sqrt{\pi y}} \sum_{n=0}^{\infty} \frac{x^n}{n!} (_H\hat{f}(y)),$$

$$_H\hat{f}(y) := \frac{1}{2\sqrt{y}} \int_{-\infty}^{\infty} e^{-\frac{\xi^2}{4y}} \left(\frac{\xi}{2y} + {}_{-\frac{1}{4|y|}}\hat{h} \right)^n f(\xi)\, d\xi\ {}_{\bar{y}}\theta_0,$$

$$y > 0,$$

$$(7.4.16)$$

where $_H\hat{f}(y)$ is the *Hermite transform* of the function $f(x)$.
65 The same point of view can be followed to introduce the *Laguerre transform*, which can be derived from the identity

$$e^{-y\,\partial_x x\partial_x} f(x) = \int_0^{\infty} e^{-\sigma} e^{-yx} C_0(x\sigma) f(\sigma)\, d(\sigma). \quad (7.4.17)$$

The use of the generating function leads therefore to the identification of the Laguerre transform $_L\hat{f}(y)$, as follows:

$$\int_0^{\infty} e^{-\sigma} e^{-yx} C_0(x\sigma) f(\sigma)\, d(\sigma) = \sum_{n=0}^{\infty} \frac{x^n}{n!}\, _L\hat{f}_n(y),$$

$$_L\hat{f}_n(y) := \int_0^{\infty} e^{-\sigma} (y - \hat{c}\sigma)^n f(\sigma)\, d\sigma\ \varphi_0.$$

$$(7.4.18)$$

Both Laguerre and Hermite transforms are linked to the orthogonal properties of these family of polynomials and the relevant definitions can be extended to higher-order polynomials or to multivariate Hermite.

66 We can combine the *Hermitian* umbra to get further generalizations, as for the three-variable third-order *HP*, which,
according to the previous formalism, can be defined as

$$H_n^{(3)}(x, y, z) = (x + {}_y\hat{h}_2 + {}_z\hat{h}_3)^n \theta_{0,y}\theta_{0,z}, \qquad (7.4.19)$$

thereby we find

$$H_n^{(3)}(x, y, z) = \sum_{s=0}^{n} \binom{n}{s} {}_{(y,z)}\hat{h}_{(2,3)}^s \, x^{n-s}\theta_{0,y}\theta_{0,z},$$

$$(7.4.20)$$

$${}_{(y,z)}\hat{h}_{(2,3)}^s = \sum_{r=0}^{s} \binom{s}{r} {}_z\hat{h}_3^{s-r} \, {}_y\hat{h}_2^r.$$

The extension of the method to bilateral generating functions
is quite straightforward, too. We consider indeed the generating function

$$G(x, y; z, w \mid t) = \sum_{n=0}^{\infty} \frac{t^n}{n!} H_n(x, y) H_n(z, w)$$

$$= \sum_{n=0}^{\infty} \frac{t^n}{n!} \left(x + {}_y\hat{h}\right)^n H_n(z, w)\theta_0$$

$$= e^{\left(x + {}_y\hat{h}\right)zt + \left[\left(x + {}_y\hat{h}\right)t\right]^2 w}\theta_0. \qquad (7.4.21)$$

The use of our technique yields

$$G(x, y; z, w \mid t) = \frac{1}{\sqrt{1 - 4yt^2w}} e^{\frac{\left(x^2w + yz^2\right)t^2 + xtz}{1 - 4yt^2w}}. \qquad (7.4.22)$$

Exotic generating functions involving, e.g., products of *LP*
and *HP* can also be obtained.

7.5 Operator Ordering and Laguerre Umbral Operators

In Chapter 1, we have touched upon the problem of *operator ordering*
arising in the solution of PDE when *umbral exponential functions
(UEF)* containing a sum of non-commuting operators are involved.
We have also argued that the problem of finding appropriate ordering
forms is, in these cases, complicated by the fact that *UEF*s do not

possess the semi-group property of the ordinary exponential function (as it has been noted in Eq. (7.10.1) and in Section 5.2).

The problem of merging ordering procedures and the umbral formalism has been introduced in [18] and further elaborated in [66, 70]. The breakthrough of the procedure may be shown using a fairly straightforward exercise, provided by a *Laguerre-derivative-based PDE* [18].

67　Let, $\forall x, \alpha, \beta \in \mathbb{R}, \forall t \in \mathbb{R}_0^+$,

$$\begin{cases} {}_l\partial_t F(x,t) = -(\alpha\, x - \beta\, \partial_x) F(x,t), \\ F(x,0) = f(x), \end{cases} \tag{7.5.1}$$

where we indicate with ${}_l\partial_t := \partial_t t\, \partial_t$ the Laguerre derivative (l-derivative) (2.6.8). We call the property[n] ${}_l\partial_t^n = \partial_t^n t^n \partial_t^n$, $\forall n \in \mathbb{R}$ in Eq. (2.6.13) and we naturally introduce the *Laguerre exponential* (*l-exponential*) (a "compromise" between an ordinary exponential and the Laguerre function) [18, 74, 83]

$$ {}_l e(\eta) := \sum_{r=0}^{\infty} \frac{\eta^r}{(r!)^2}, \quad \forall \eta \in \mathbb{R}, \tag{7.5.2}$$

or in umbral version (by using (1.1.6))

$$ {}_l e(x) = \sum_{r=0}^{\infty} \frac{(x)^r}{(r!)^2} = e^{\hat{c}x}\varphi_0. \tag{7.5.3}$$

The function ${}_l e(\eta)$ is a zeroth-order Bessel–Tricomi function[o] [166] and satisfies the "Laguerre"-eigenvalue equation $\forall \lambda \in \mathbb{R}$

$$ {}_l\partial_t \left({}_l e(\lambda\, t) \right) = \lambda \left({}_l e(\lambda\, t) \right), \tag{7.5.4}$$

---

[n]Albeit we are not making any explicit reference to fractional derivatives, we quote this example to underscore the generality of the procedure we envisage, whose operator nature allows a fairly straightforward extension to the fractional case.
[o]It can be expressed in terms of the zeroth-order modified Bessel function (see Chapter 4) through the identity ${}_l e(\eta) = I_0(2\sqrt{\eta})$.

indeed, since l-derivative satisfies ordinary series integration theorem, we get

$$\imath\partial_t\left(\imath e(\lambda\,t)\right) = \imath\partial_t\sum_{r=0}^{\infty}\frac{(\lambda\,t)^r}{r!^2} = \sum_{r=0}^{\infty}\imath\partial_t\frac{(\lambda\,t)^r}{r!^2} = \lambda\sum_{r=1}^{\infty}\frac{(\lambda\,t)^{r-1}}{(r-1)!^2}$$

$$= \lambda\imath e(\lambda\,t),\tag{7.5.5}$$

(which corroborates the interpretation of its role as that of a *l*-exponential) and furthermore

$$e^{\kappa\,\imath\partial_t}\frac{t^n}{n!} = \sum_{r=0}^{\infty}\frac{\kappa^r}{r!}\imath\partial_t^r\frac{t^n}{n!} = \sum_{r=0}^{\infty}\frac{\kappa^r}{r!}\partial_t^r t^r\partial_t^r\frac{t^n}{n!} = \sum_{r=0}^{\infty}\binom{n}{r}\frac{\kappa^r x^{n-r}}{(n-r)!}$$

$$\tag{7.5.6}$$

obtained after expanding the *l*-exponential and by applying the property of the *l*-derivative operator (2.6.13). The solution of Eq. (7.5.1) then can be written using the evolution operator formalism in which the exponential is replaced by

$$F(x,t) = \imath e\left(-t\left(\alpha\,x - \beta\,\partial_x\right)\right)\tag{7.5.7}$$

or, in umbral form, as

$$F(x,t) = e^{-\hat{c}\,t\,(\alpha\,x-\beta\,\partial_x)}f(x)\varphi_0.\tag{7.5.8}$$

The (pseudo) exponential evolution operator cannot be disentangled into the product of two exponentials, because the operators in the argument of the exponential are not commuting. However, by applying the Weyl ordering rule (see application (7.10.1)), we obtain

$$F(x,t) = e^{-\frac{(\hat{c}t)^2}{2}\alpha\beta}e^{-\hat{c}\,t\,\alpha\,x}e^{\hat{c}\beta\,t\,\partial_x}f(x)\varphi_0$$

$$= e^{-\frac{(\hat{c}t)^2}{2}\alpha\beta}e^{-\hat{c}\,t\,\alpha\,x}f(x + \hat{c}\beta\,t)\varphi_0.\tag{7.5.9}$$

It is important to note that the operational ordering brings into play a further term depending on the square of the umbral operator $\hat{c}$, which commutes with the differential operators x, ∂_x. Assuming for simplicity $f(x) = 1$, we find that

$$F(x,t) = e^{-\frac{(\hat{c}t)^2}{2}\alpha\beta}e^{-\hat{c}\,t\,\alpha\,x}\varphi_0.\tag{7.5.10}$$

Remark 5. The example we have discussed is sufficient to show how the umbral formalism naturally yields the solution of evolution problems involving Laguerre derivative and non-commuting operators. Albeit Eq. (7.5.10) is just providing the solution for a mathematical problem (without any particular meaning except that of being associated with a Laguerre evolution problem) it is worth speculating a little more on it. After expanding the exponential containing x, we find

$$F(x,t) = \sum_{n=0}^{\infty} \frac{(\hat{c}t\alpha x)^n}{n!} e^{-\frac{(\hat{c}t)^2}{2}\alpha\beta} \varphi_0 = \sum_{n=0}^{\infty} \frac{(\alpha\,tx)^n}{n!} {}_l e_n^{(2)}\left(-\frac{\alpha\beta\,t^2}{2}\right),$$

(7.5.11)

where ${}_l e_n^{(m)}(x)$ is the Bessel-like function so defined [18, 74, 83]

$$ {}_l e_n^{(m)}(x) = \sum_{r=0}^{\infty} \frac{x^r}{r!\,\Gamma(m\,r+n+1)}. $$

(7.5.12)

The results obtained so far have extended to the Laguerre derivative case, analogous conclusions contained in [66, 70] where, among other things, the fractional Poisson distribution has been exploited, within the context of the definition of a new set of coherent states generated in the study of a Fractional Schrödinger equation, driving the process of one photon emission-absorption dynamics (see Section 109).

7.6   Ramanujan Master Theorem

We have used the umbral methods to evaluate integrals and we have obtained a framing of the procedure within an adequate mathematical context. The results we have obtained trace however back to the so-called *Ramanujan Master Theorem (RMT)* [13, 25, 102], which is illustrated in this subsection. The theorem can be stated as follows.

Theorem 6 (RMT). *If the function $f(x)$ satisfies the series expansion*

$$\mathbf{f(x)} = \sum_{n=0}^{\infty} \mathbf{f_n} \frac{(-1)^n x^n}{n!}$$

(7.6.1)

in a neighborhood of the origin, with $f_0 = \varphi_0$, then

$$\int_0^\infty x^{\nu-1} f(x)\, dx = \Gamma(\nu) f_{-\nu}. \tag{7.6.2}$$

The use of the umbral formalism (see, e.g., the property (1.1.6)) allows indeed to write

$$f(x) = e^{-\hat{c}\,x}\, f_0, \tag{7.6.3}$$

since $\hat{c}^n\, f_0 := f_n$ and therefore the integral (7.6.2) can be given in the form

$$\int_0^\infty x^{\nu-1} e^{-\hat{c}x}\, dx\, f_0 = \Gamma(\nu)\hat{c}^{-\nu} f_0 = \Gamma(\nu) f_{-\nu}. \tag{7.6.4}$$

68 The same procedure can be used to prove

$$\int_0^\infty x^{\nu-2} f(x)\,dx = \frac{1}{m}\Gamma\left(\frac{\nu-1}{m}\right) f_{\frac{1-\nu}{m}},$$

$$f(x) = \sum_{n=0}^\infty f_n \frac{(-1)^n x^{mn}}{n!}, \tag{7.6.5}$$

obtained after noting that

$$\int_0^\infty x^{\nu-2} e^{-\hat{c}\,x^m}\, dx\, f_0 = \frac{1}{m}\int_0^\infty t^{\left(\frac{\nu-1}{m}-1\right)} e^{-\hat{c}\,t}\, dt\, f_0$$

$$= \frac{1}{m}\Gamma\left(\frac{\nu-1}{m}\right) f_{-\frac{\nu-1}{m}}. \tag{7.6.6}$$

69 We can use, e.g., the *RMT* to prove the identities

$$\int_0^\infty x^{\nu-1} C_\alpha(x)\,dx = \frac{\Gamma(\nu)}{\Gamma(\alpha-\nu+1)}, \tag{7.6.7}$$

where $C_\alpha(x)$ is the Tricomi–Bessel function of order α (see (1.1.2)) which in umbral form (3.1.40) can be written as $C_\nu(x) = \hat{c}^\nu e^{-\hat{c}x}\varphi_0$. We note in fact that, according to the umbral formalism,

$$\int_0^\infty x^{\nu-1} C_\alpha(x) dx = \hat{c}^\alpha \int_0^\infty x^{\nu-1} e^{-\hat{c}x} dx \, \varphi_0 = \Gamma(\nu) \hat{c}^{\alpha-\nu} \varphi_0$$

$$= \frac{\Gamma(\nu)}{\Gamma(\alpha - \nu + 1)}. \tag{7.6.8}$$

7.7   Products of Bessel Functions and Associated Polynomials

Let

$$f(x; a, b) := J_0(ax) J_0(bx) \tag{7.7.1}$$

a product of zeroth-order Bessel functions, which can be formally written as the product of two Gaussians (4.1.2), namely[P]

$$f(x; a, b) = e^{-(a^2 \hat{c}_1 + b^2 \hat{c}_2)\left(\frac{x}{2}\right)^2} \varphi_0^{(1)} \varphi_0^{(2)}, \tag{7.7.2}$$

where $\varphi_0^{(\alpha)}$ are the umbral vacua on which the operators $\hat{c}_\alpha$ act. The series expansion of the exponential and the use of the previously outlined rules yield

$$f(x; a, b) = \sum_{r=0}^\infty \frac{(-1)^r}{r!} l_r(a^2, b^2) \left(\frac{x}{2}\right)^{2r},$$

$$l_r(a, b) = r! \sum_{s=0}^r \frac{a^{(r-s)} b^s}{(s!)^2 \left[(r-s)!\right]^2}. \tag{7.7.3}$$

Leaving for the moment unspecified the nature of the polynomials $l_r(a, b)$, we note that the function $f(x; a, b)$ can be cast in the umbral form

$$f(x; a, b) = e^{-\hat{l}\left(\frac{x}{2}\right)^2} \Phi_0, \qquad \hat{l}^\nu \Phi_0 = l_\nu(a^2, b^2). \tag{7.7.4}$$

The action of the operator $\hat{l}$ on the corresponding umbral vacuum holds for any real (positive/negative) or complex value of the

---

[P]Even though not explicitly stated, it is evident that in the present formalism we have

$$[J_0(x)]^2 = e^{-\hat{c}\left(\frac{x}{2}\right)^2} e^{-\hat{c}\left(\frac{x}{2}\right)^2} \varphi_0.$$

exponent ν. We have concluded that the product of two cylindrical Bessels is the umbral equivalent of a BF and thus the umbra of a Gaussian.

70 Such a conclusion turns particularly useful if we are interested in the evaluation of the integrals of the function $f(x; a, b)$, a straightforward use of the so far developed procedure yields $\forall a, b \in \mathbb{R} :\mid a \mid > \mid b \mid$, we get

$$\int_{-\infty}^{+\infty} f(x; a, b)dx = \int_{-\infty}^{+\infty} e^{-\hat{l}\left(\frac{x}{2}\right)^2} dx \, \Phi_0 = 2\sqrt{\pi} \, \hat{l}^{-\frac{1}{2}} \Phi_0$$

$$= 2\sqrt{\pi} \, l_{-\frac{1}{2}}(a^2, b^2),$$

$$l_{-\frac{1}{2}}(a^2, b^2) = \Gamma\left(\frac{1}{2}\right) \sum_{s=0}^{\infty} \frac{a^{-2\left(\frac{1}{2}+s\right)} b^{2s}}{s!^2 \, \Gamma\left(\frac{1}{2}-s\right)^2} = \frac{1}{\sqrt{\pi} \mid a \mid} K\left(\frac{b}{a}\right),$$

$$K(k) = \,_2F_1\left(\frac{1}{2}, \frac{1}{2}; 1; k^2\right) = \sum_{s=0}^{\infty} \left[\frac{(2s)!}{2^{2s} \, s!^2}\right]^2 k^{2s},$$

$$(7.7.5)$$

where $\,_2F_1(a, b; c; z)$ is the confluent hypergeometric function [2].

71 We have left open the question on the nature of the polynomials $l_r(a, b)$. Although we will discuss more deeply this point in the forthcoming sections, here we note that they can be viewed as a particular case of the Jacobi polynomials (seen in Section (3.1.2)), as it can be inferred from the identity [18]:

$$l_r\left(\frac{x-1}{2}, \frac{x+1}{2}\right) = \frac{1}{r!} P_r^{(0,0)}(x),$$

$$P_n^{(\alpha,\beta)}(x) = \sum_{s=0}^{n} \binom{n+\alpha}{s} \binom{n+\beta}{n-s} \left(\frac{x-1}{2}\right)^{n-s} \left(\frac{x+1}{2}\right)^s.$$

$$(7.7.6)$$

72 By the use of Gaussian umbral form of real-order Bessel function (4.2.1), we obtain the following general expression for the product of two cylindrical Bessel functions of order ν, μ, respectively,

$$
\begin{aligned}
f_{\nu,\mu}(x;a,b) &= J_\nu(ax)J_\mu(bx) \\
&= \left(\frac{x}{2}\right)^{\nu+\mu}(a\hat{c}_1)^\nu(b\hat{c}_2)^\mu e^{-(a^2\hat{c}_1+b^2\hat{c}_2)\left(\frac{x}{2}\right)^2}\varphi_0^{(1)}\varphi_0^{(2)} \\
&= \left(\frac{x}{2}\right)^{\nu+\mu}a^\nu b^\mu \sum_{r=0}^\infty \frac{(-1)^r}{r!}l_r^{(\nu,\mu)}(a^2,b^2)\left(\frac{x}{2}\right)^{2r},
\end{aligned}
$$

$$
l_r^{(\nu,\mu)}(a,b) = r!\sum_{s=0}^r \frac{a^{(r-s)}b^s}{\Gamma(\mu+s+1)\Gamma(\nu+r-s+1)s!(r-s)!}.
$$

$$(7.7.7)$$

73 The product of three zeroth order Bessel functions can be written as

$$
f(x;a_1,a_2,a_3) = e^{-(a_1^2\hat{c}_1+a_2^2\hat{c}_2+a_3^2\hat{c}_3)\left(\frac{x}{2}\right)^2}\varphi_0^{(1)}\varphi_0^{(2)}\varphi_0^{(3)} \quad (7.7.8)
$$

or, in explicit form,

$$
f(x;a_1,a_2,a_3) = \sum_{r=0}^\infty \frac{(-1)^r}{r!}l_r(a_1^2,a_2^2,a_3^2)\left(\frac{x}{2}\right)^{2r},
$$

$$
l_r(x_1,x_2,x_3) = r!\sum_{s=0}^r \frac{x_3^{(r-s)}}{s![(r-s)!]^2}l_s(x_1,x_2).
$$

$$(7.7.9)$$

It is evident that the extension to the case of n Bessel functions writes as in the first of Eqs. (7.7.9) with

$$
l_r(x_1,\ldots,x_n) = r!\sum_{s=0}^r \frac{x_n^{(r-s)}}{s![(r-s)!]^2}l_s(x_1,\ldots,x_{n-1}). \quad (7.7.10)
$$

In [127], all the α parameters (actually the variables of the l_r polynomials) are assumed to be 1.

74 From a formal point of view, the use of the multinomial expansion allows to define the previous family of polynomials as

$$l_r(x_1, \ldots, x_n) = (x_1 \hat{c}_1 + \cdots + x_n \hat{c}_n)^r \varphi_0^{(1)} \ldots \varphi_0^{(n)} \quad (7.7.11)$$

and, the use of the multinomial expansion yields

$$l_r(x_1, \ldots, x_n) = \sum_{k_1 + \cdots + k_n = r} \binom{r}{k_1 \ldots k_r} \frac{x_1^{k_1}}{(k_1!)^2} \cdots \frac{x_n^{k_n}}{(k_n!)^2}.$$
$$(7.7.12)$$

75 Going back to the two-variable case, it is easy to check that they satisfy the differential equation

$$\partial_{x_1} x_1 \partial_{x_1} l_r(x_1, x_2) = \partial_{x_2} x_2 \partial_{x_2} l_r(x_1, x_2) = r l_{r-1}(x_1, x_2),$$
$$(7.7.13)$$

with $\partial_x x \partial_x$ l-derivative (2.6.8), recalling Eq. (2.7.1).

76 To obtain the extension to the product of arbitrary cylindrical Bessel, it will be sufficient to replace in the previous equations the function $l_r(a_1^2, \ldots, a_n^2)$ with $l_r^{(\nu_1, \ldots, \nu_n)}(a_1^2, \ldots, a_n^2)$

$$l_r^{(\nu_1, \ldots, \nu_n)}(x_1, \ldots, x_n)$$
$$= \hat{c}_1^{\nu_1} \ldots \hat{c}_n^{\nu_n} (x_1 \hat{c}_1 + \cdots + x_n \hat{c}_n)^r \varphi_0^{(1)} \ldots \varphi_0^{(n)}$$
$$= \sum_{k_1 + \cdots + k_n = r} \binom{r}{k_1 \ldots k_r} \frac{x_1^{k_1}}{k_1! \Gamma(\nu_1 + k_1 + 1)} \quad (7.7.14)$$
$$\cdots \frac{x_n^{k_n}}{k_n! \Gamma(\nu_n + k_n + 1)}$$

thus getting an expression closely similar to that derived by Brychkov in [34]

$$\prod_{s=1}^{n} J_{\nu_s}(a_s x)$$
$$= \left(\frac{x}{2}\right)^{\sum_{s=1}^{n} \nu_s} \left(\prod_{k=1}^{n} a_k^{\nu_k}\right) \sum_{r=0}^{\infty} \frac{(-1)^r}{r!} l_r^{(\nu_1, \ldots, \nu_n)} \quad (7.7.15)$$
$$(a_1^2, \ldots, a_n^2) \left(\frac{x}{2}\right)^{2r}.$$

Proposition 48. $\forall x, \nu \in \mathbb{R}$

$$\hat{l}_\nu(x) = \sum_{r=0}^{\infty} \frac{\Gamma(\nu+1)}{r!\Gamma(r+\nu+1)} \left(\frac{x}{2}\right)^{2r} = \Gamma(\nu+1)\hat{c}^\nu e^{\hat{c}\left(\frac{x}{2}\right)^2} \varphi_0. \quad (7.7.16)$$

According to our formalism, the relevant kth power reads

$$(\hat{l}_\nu(x))^k = \Gamma(\nu+1)^k \sum_{r=0}^{\infty} \frac{1}{r!} l_r^{(\nu,\dots,\nu)}(1,\dots,1) \left(\frac{x}{2}\right)^{2r}. \quad (7.7.17)$$

The polynomials defined in [127] are expressible in terms of our $l_r^{(\nu_1,\dots,\nu_n)}(x_1,\dots,x_n)$ *as*

$$B_r^{(\nu)}(k) = \Gamma(\nu+1)^{k-1}\Gamma(r+\nu+1)l_r^{\{\nu\}}(k),$$
$$l_r^{(\nu,\dots,\nu)}(1,\dots,1) = l_r^{\{\nu\}}(k),$$
$$l_{r+1}^{\{\nu\}}(k) = \sum_{j=1}^{k} l_r^{\{\nu+1_j\}}(k), \quad \{\nu+1_j,k\} = (\nu,\dots,\nu+1,\dots,\nu).$$

$$(7.7.18)$$

In the forthcoming section, we discuss the nature of the polynomials $l_r^{\{\nu\}}(k)$.

7.7.1 $l_r^{\{\nu\}}(k)$ *Polynomials*

We observe that the relevant generating function of $l_r^{\{\nu\}}(k)$ polynomials is expressible in terms of product of Bessel-like functions, namely

Proposition 49. $\forall t \in \mathbb{R}$

$$\sum_{r=0}^{\infty} \frac{(-t)^r}{r!} l_r^{(\nu_1,\dots,\nu_n)}(x_1,\dots,x_n) = \prod_{j=1}^{n} C_{\nu_j}(tx_j), \quad (7.7.19)$$

where $C_\nu(x)$ *denotes the Tricomi–Bessel function[q] of order* ν (3.1.40).

---

[q]We recall the link between Tricomi–Bessel and Bessel functions (3.1.41) $C_\nu(x) = x^{-\frac{\nu}{2}} J_\nu(2\sqrt{x})$.

Proposition 50. *By using Eq. (7.7.14), we find that*

$$l_r^{\{\nu\}}(k+1)$$

$$= \hat{c}_{k+1}^{\nu}\hat{c}_1^{\nu}\ldots\hat{c}_k^{\nu}(\hat{c}_1 + \cdots + \hat{c}_k + \hat{c}_{k+1})^r \varphi_0^{(1)}\ldots\varphi_0^{(k)}\varphi_0^{(k+1)}$$

$$= \hat{c}_{k+1}^{\nu}\sum_{j=0}^{r}\binom{r}{j}\hat{c}_{k+1}^{r-j}l_j^{\nu}(k)\varphi_0^{(k+1)}$$

$$= \sum_{j=0}^{r}\binom{r}{j}\frac{1}{\Gamma(r-j+\nu+1)}l_j^{\nu}(k). \tag{7.7.20}$$

The various identities reported in [127] *follow from the above equation, which can be generalized in various ways, as, e.g.,*

$$l_r^{\{\nu\}}(k+s) = \hat{c}_{k+1}^{\nu}\ldots\hat{c}_{k+s}^{\nu}\hat{c}_1^{\nu}\ldots\hat{c}_k^{\nu}(\hat{c}_1 + \cdots + \hat{c}_k + \hat{c}_{k+1} + \ldots$$

$$+ \cdots + \hat{c}_{k+s})^r \varphi_0^{(1)}\ldots\varphi_0^{(k)}\varphi_0^{(k+1)}\ldots\varphi_0^{(k+s)}$$

$$= \sum_{j=0}^{r}\binom{r}{j}l_{r-j}^{\{\nu\}}(s)l_j^{\{\nu\}}(k). \tag{7.7.21}$$

We have noted in Eq. (7.7.5) that the use of straightforward algebraic manipulations allows the derivation of an expression yielding the integral of the product of two cylindrical Bessel functions. We have checked that the extension to the products of three or more is anyway feasible. Regarding the case of an integral of the product of three Bessel functions then, we find what follows.

77 If $\mid a_3 \mid > \mid a_2 \mid > \mid a_1 \mid$, we find that

$$\int_{-\infty}^{+\infty} f(x; a_1, a_2, a_3)dx = 2\sqrt{\pi}l_{-\frac{1}{2}}(a_1^2, a_2^2, a_3^2),$$

$$l_{-\frac{1}{2}}(a_1, a_2, a_3) = \Gamma\left(\frac{1}{2}\right)\sum_{s=0}^{\infty}\frac{a_3^{-\left(\frac{1}{2}+s\right)}}{s!\,\Gamma\left(\frac{1}{2}-s\right)^2}l_s(a_1, a_2). \tag{7.7.22}$$

In Eq. (7.7.5), we have recognized that the series defining $l_{-\frac{1}{2}}(a,b)$ can be recognized as that defining a quarter period elliptic integral, in this case we obtain

$$l_{-\frac{1}{2}}(a_1,a_2,a_3) = \frac{1}{\sqrt{\pi \mid a_3 \mid}} F(a_1,a_2,a_3),$$

$$F(a_1,a_2,a_3) = \sum_{s=0}^{\infty} \left[\frac{(2s)!}{2^{2s}(s!)^2} \right]^2 \frac{l_s(a_1,a_2)}{a_3^s} = {}_2F_1\left(\frac{1}{2}, \frac{1}{2}; 1; \frac{\tilde{f}}{a_3} \right) \chi_0,$$

$$\hat{f}^r \chi_0 = s! l_s(a_1,a_2),$$

$$(7.7.23)$$

namely, we have reduced the series at least formally to the same hypergeometric defining the elliptic integral period. This result can be easily generalized to the case of an arbitrary product.

Observation 16. A further element of interest concerns the fact that, since, as already remarked, by replacing $\hat{f}$ with $\hat{c}$ the functions defining the product of Bessel and the Bessel functions are umbral equivalent, we can take advantage from the formalism to establish, e.g., the *n-th* derivative of the $f(x;a,b)$ functions.

Proposition 51. *By noting again that it is formally written as a Gaussian, by the use of property* $\hat{D}_x^n e^{ax^2} = H_n(2ax,a)e^{ax^2}$ *in (1.2.7), we can write the nth derivative of the product of two Bessel functions in terms of the two-variable HP as*

$$\hat{D}_x^n f(x;a,b) = \hat{D}_x^n e^{-\hat{l}\left(\frac{x}{2}\right)^2} \Phi_0$$

$$= H_n\left(-\hat{l}\,\frac{x}{2}, -\frac{\hat{l}}{4} \right) e^{-\hat{l}\left(\frac{x}{2}\right)^2} \Phi_0$$

$$= (-1)^n H_n\left(\hat{l}\,\frac{x}{2}, -\frac{\hat{l}}{4} \right) e^{-\hat{l}\left(\frac{x}{2}\right)^2} \Phi_0. \quad (7.7.24)$$

The use of the properties of the $\hat{l}$-operator finally yields the explicit result as

$$\hat{D}_x^n f(x; a, b) = \frac{(-1)^n}{2^n} n! \sum_{r=0}^{\lfloor \frac{n}{2} \rfloor} \frac{(-1)^r x^{n-2r}}{r!(n-2r)!} \,_{(n-r)} f(x; a, b),$$

$$_{(s)} f(x; a, b) = \sum_{k=0}^{\infty} \frac{(-1)^k}{k!} l_{k+s}(a^2, b^2) \left(\frac{x}{2}\right)^{2k}.$$

(7.7.25)

78 Now we just touch upon the application of the formalism to the theory of multi-index Bessel functions. We recall that the *Humbert functions*[r] [15] within the present formalism are defined as

$$I_{m_1, m_2}(x) = \hat{c}_1^{m_1} \hat{c}_2^{m_2} e^{\hat{c}_1 \hat{c}_2 x} \varphi_0^{(1)}(0) \varphi_0^{(2)}(0)$$

$$= \sum_{s=0}^{\infty} \frac{x^r}{r!(m_1 + r)!(m_2 + r)!}.$$

(7.7.26)

The relevant properties are easily deduced, for example, we find

$$\hat{D}_x I_{m_1, m_2}(x) = \hat{c}_1^{m_1+1} \hat{c}_2^{m_2+1} e^{\hat{c}_1 \hat{c}_2 x} \varphi_0^{(1)}(0) \varphi_0^{(2)}(0)$$

$$= I_{m_1+1, m_2+1}(x)$$

(7.7.27)

or, by applying the same integration procedure as before, we obtain

$$\int_{-\infty}^{+\infty} I_{0,0}(x) e^{-\beta x^2} dx = \sqrt{\frac{\pi}{\beta}} I_{0,0}\left(\frac{1}{4\beta} \mid 2\right),$$

$$I_{m_1, m_2}(x \mid k) = \sum_{r=0}^{\infty} \frac{x^r}{r! \Gamma(kr + 1 + m_1) \Gamma(kr + 1 + m_2)}.$$

(7.7.28)

The second of Eq. (7.7.28) is a *two indexes Bessel–Wright function* (1.1.1) and the Gaussian integral in the first of

---

[r]With respect to the definition of Humbert–Bessel–like functions in (4.4.17), we used here $I_{m_1, m_2}(x) = J_{m_1, m_2}(-x)$.

Eq. (7.7.28) can be viewed as the integral transform adopted for their definition.

In this section, we have shown that our formalism of umbral nature can be exploited to simplify in a significant way the technicalities underlying the theory of Bessel functions and of their manipulations leading to combinations or to the introduction of new forms.

7.8   Mittag–Leffler Functions and Fractional Calculus

The Mittag–Leffler functions plays a pivotal role within the context of the fractional calculus, namely in the branch of analysis using derivative and integral operators, which can be interpreted as derivative operators of (positive or negative) fractional order. Here we provide a flavor of how the relevant formalism and umbral operational methods can be merged.

79 Consider the exponential umbral image of the Mittag–Leffler function in (1.3.14) $E_{\alpha,\beta}(x) = e^{\alpha,\beta \hat{d}x}\psi_0$ with the umbral operator introduced in (1.3.13) and proved in Proposition 4. Prove that the successive derivatives of the function (we have indicated in Eq. (7.2.14) $E_{\alpha,1}(x) := E_\alpha(x)$) are

$$\partial_x^m E_\alpha(x) = \sum_{n=0}^{\infty} \frac{(n+m)!}{n!} \frac{x^n}{\Gamma(\alpha(n+m))+1}, \quad \forall m \in \mathbb{N}.$$

(7.8.1)

Hint: By keeping the mth derivative of both sides of Eq. (1.3.14), we find

$$E_\alpha^{(m)}(x) = {}_\alpha\hat{d}^m e^{\alpha\hat{d}x}\psi_0 = \sum_{n=0}^{\infty} \frac{{}_\alpha\hat{d}^{n+m}}{n!} x^n \psi_0 \cdots$$

80 Find the same result of the previous exercise using the rational image in Eq. (1.3.2) and already used in Exercise (43).

Hint: Note that

$$\partial_x^m E_\alpha(x) = \partial_x^m \int_0^\infty e^{-s} e^{s\hat{c}^\alpha x} ds \, \varphi_0$$

$$= c^{\hat{\alpha} m} \, \partial_x^m \int_0^\infty e^{-s} s^m e^{s\hat{c}^\alpha x} ds \, \varphi_0. \quad (7.8.2)$$

81 Extend the previous results to $E_{\alpha,\beta}(x)$.

82 Review Examples 8 and 9 in Chapter 1 and show that

$$I_{ML}(\alpha,\kappa) = \int_{-\infty}^\infty e^{-x^2} E_{\alpha,1}(ikx) dx$$

$$= \sqrt{\pi} \int_0^\infty W_0^{2\alpha} \left(-\left(\frac{ks}{2}\right)^2 \right) ds. \quad (7.8.3)$$

Hint:
$\int_{-\infty}^\infty e^{-x^2} E_{\alpha,1}(ikx) dx = \int_{-\infty}^\infty e^{-x^2} \left(\int_0^\infty e^{-s} e^{i\hat{c}^\alpha kxs} ds \right) dx \, \varphi_0 = \ldots$

83 Show that

$$\int_0^x \xi^m E_{\alpha,1}(\xi) d\xi = \sum_{s=0}^\infty \frac{(-1)^s m!}{(m+s+1)!} x^{m+s+1} E_{\alpha,1}^{(s)}(x). \quad (7.8.4)$$

Hint: Use the repeated integration by part.

84 Define the pseudo-Poisson distribution

$$P_\mu(x,n) = \frac{x^n}{n!} E_\mu^{(n)}(z) \mid_{z=-x} \quad (7.8.5)$$

and prove that it reduces to a standard Poisson distribution for $\mu = 1$. Discuss whether (7.8.5) can be considered a probability distribution (used in Chapters 1 and 2 in fractional form, too).

Hint: The first question is straightforward and we find

$$P_1(x,n) = \frac{x^n}{n!} e^{-x}. \quad (7.8.6)$$

Regarding the second, we note that

$$P_\mu(x,n) = \frac{x^n}{n!} \mu \hat{d}^n e^{-\mu \hat{d}x} \psi_0, \quad (7.8.7)$$

it is easily checked that

$$\sum_{n=0}^{\infty} P_\mu(x,n) = 1. \tag{7.8.8}$$

85 Show that

$$\overline{n} = \sum_{n=0}^{\infty} n\, P_\mu(x,n) = \frac{x}{\Gamma(\mu+1)} \tag{7.8.9}$$

and

$$\overline{n^2} = \sum_{n=0}^{\infty} n^2\, P_\mu(x,n) = \frac{3x^2}{\Gamma(2\mu+1)} + \overline{n}. \tag{7.8.10}$$

Lemma 15. *Let $E_{n,1}(x)$ be the ML function, $\forall n \in \mathbb{N}$, $\forall x \in \mathbb{R}$ and let $\lambda \in \mathbb{R} \Rightarrow E_{n,1}(x)$ be eigenfunction of ∂_x^n.*

$$\partial_x^n E_{n,1}(\lambda x^n) = \lambda E_{n,1}(\lambda x^n). \tag{7.8.11}$$

Proof. $\forall n \in \mathbb{N}$, $\forall x, \lambda \in \mathbb{R}$ [66]

$$\partial_x^n E_{n,1}(\lambda x^n) = \partial_x^n \sum_{r=0}^{\infty} \frac{\lambda^r x^{nr}}{\Gamma(nr+1)} = \sum_{r=0}^{\infty} \partial_x^n \frac{\lambda^r x^{nr}}{\Gamma(nr+1)}$$

$$= \sum_{r=0}^{\infty} \lambda^r \frac{(nr)!}{(nr-n)!} \frac{x^{nr-n}}{\Gamma(nr+1)}$$

$$= \lambda \sum_{r=1}^{\infty} \frac{\lambda^{r-1} x^{n(r-1)}}{(n(r-1))!} = \lambda E_{n,1}(\lambda x^n). \tag{7.8.12}$$

We have interchanged the order derivative and summation in view of the uniform convergence of the series defining the *ML* function. $\square$

An analogous identity can be extended also to the case of real order *ML* functions. In this case, *derivatives of non-integer order should be considered.*

Corollary 49. *By using the Euler–Riemann–Liouville definition* [134] *for real order derivative* [66]

$$\partial_x^\alpha x^\nu = \frac{\Gamma(\nu+1)}{\Gamma(\nu-\alpha+1)} x^{\nu-\alpha}, \qquad \forall x, \alpha, \nu \in \mathbb{R}, \tag{7.8.13}$$

we find[s] that the ML function, $E_{\alpha,1}(\lambda x^\alpha)$, is an eigenfunction of the ∂_x^α operator $\forall \alpha \in \mathbb{R}^+$

$$\partial_x^\alpha \, E_{\alpha,1}(\lambda \, x^\alpha) = \lambda \, E_{\alpha,1}(\lambda x^\alpha) + \frac{x^{-\alpha}}{\Gamma(1-\alpha)}. \tag{7.8.14}$$

Proof. $\forall x, \lambda \in \mathbb{R}$, $\forall \alpha \in \mathbb{R}^+$

$$\partial_x^\alpha \, E_{\alpha,1}(\lambda \, x^\alpha) = \partial_x^\alpha \, 1 + \partial_x^\alpha \sum_{r=1}^{\infty} \frac{(\lambda x^\alpha)^r}{\Gamma(1+\alpha r)}$$

$$= \frac{x^{-\alpha}}{\Gamma(1-\alpha)} + \sum_{r=1}^{\infty} \partial_x^\alpha \, \frac{(\lambda x^\alpha)^r}{\Gamma(1+\alpha r)}$$

$$= \lambda \, E_{\alpha,1}(\lambda x^\alpha) + \frac{x^{-\alpha}}{\Gamma(1-\alpha)}. \tag{7.8.15}$$

The interchange between the summation and derivative is ensured by the uniform convergence of the series for $\alpha > 0$. The last term is due to the fact that the fractional derivative in the sense of the Riemann–Liouville (7.8.13) acts on a costant and does not vanish.□

This result can be used for various kinds of applications in different fields, as, e.g., for the solution of the following *fractional evolution problem* [100].

86 Proof of Eq. (1.3.11) $I_{\alpha,\beta} = \int_{-\infty}^{\infty} E_{\alpha,\beta}(-x^2)dx = \dfrac{\pi}{\Gamma\left(\beta - \frac{\alpha}{2}\right)}.$

$$I_{\alpha,\,\beta} = \int_{-\infty}^{\infty} E_{\alpha,\beta}(-x^2)dx = \left(\sqrt{\pi}\hat{c}^{\beta-\frac{\alpha}{2}-1} \int_{0}^{\infty} e^{-s}s^{-\frac{1}{2}}ds\right)\varphi_0$$

$$= \sqrt{\pi} \, \Gamma\left(\frac{1}{2}\right) \hat{c}^{\beta-\frac{\alpha}{2}-1}\varphi_0$$

$$= \frac{\pi}{\Gamma\left(\beta - \frac{\alpha}{2}\right)}, \quad \forall \alpha, \beta \in \mathbb{R}^+. \tag{7.8.16}$$

---

[s]The extra term emerges because, according to Eq. (7.8.13), the fractional derivative of a constant does not vanish.

87 The *ML* Gaussian-like integral $\forall \alpha, \beta \in \mathbb{R}^+$ can, accordingly (by the use of Eqs. (1.3.7), (1.1.13), (1.2.2), (1.0.1), (1.0.4) and (1.1.6)), be written as

$$
\begin{aligned}
I_{\alpha,\beta} &= \int_{-\infty}^{\infty} E_{\alpha,\beta}(-x^2)dx = \int_{-\infty}^{\infty}\int_0^{\infty} e^{-s}W_{\beta-1}^{\alpha}(-x^2 s)ds\,dx \\
&= \int_{-\infty}^{\infty}\int_0^{\infty} e^{-s}\hat{c}^{\beta-1}e^{-\hat{c}^{\alpha}x^2 s}\varphi_0\,ds\,dx \\
&= \hat{c}^{\beta-1}\int_{-\infty}^{\infty}\int_0^{\infty} e^{-s}e^{-\hat{c}^{\alpha}x^2 s}ds\,dx\,\varphi_0 \\
&= \hat{c}^{\beta-1}\int_0^{\infty} e^{-s}ds\int_{-\infty}^{\infty} e^{-\hat{c}^{\alpha}s\,x^2}dx\,\varphi_0 \\
&= \hat{c}^{\beta-1}\int_0^{\infty} e^{-s}ds\,\sqrt{\frac{\pi}{\hat{c}^{\alpha}s}}\,\varphi_0 \\
&= \sqrt{\pi}\,\hat{c}^{\beta-\frac{\alpha}{2}-1}\int_0^{\infty} e^{-s}s^{-\frac{1}{2}}ds\,\varphi_0 = \sqrt{\pi}\,\Gamma\left(\frac{1}{2}\right)\hat{c}^{\beta-\frac{\alpha}{2}-1}\varphi_0 \\
&= \frac{\pi}{\Gamma\left(\beta-\frac{\alpha}{2}\right)}.
\end{aligned} \tag{7.8.17}
$$

88 The following *fractional evolution problem*, $\forall x \in \mathbb{R}, \forall \alpha \in \mathbb{R}^+, \forall t \in \mathbb{R}_0^+$:

$$
\begin{cases}
\partial_t^{\alpha}F(x,t) = \partial_x^2\,F(x,t) + \dfrac{t^{-\alpha}}{\Gamma(1-\alpha)}f(x), \\
F(x,0) = f(x),
\end{cases} \tag{7.8.18}
$$

defines a *time-fractional diffusive equation*. According to the previous discussion, to the fact that the *ML* "$E_{\alpha,1}(t^{\alpha})$" is an eigenfunction of the fractional derivative operator, and according to the definition (7.8.14) and considering the formalism developed so far, we can obtain the relevant solution in the form [66–100]

$$
F(x,t) = E_{\alpha,1}(t^{\alpha}\partial_x^2)\,f(x), \tag{7.8.19}
$$

where, for the problem under study, we have that $\forall \alpha \in \mathbb{R}^+, \forall t \in \mathbb{R}_0^+$, $\mathbf{E}_{\alpha,1}(\mathbf{t}^{\alpha}\partial_{\mathbf{x}}^2)$ is the *pseudo-evolution operator*

(PEO). The relevant action on the initial function can be espressed as [66]

$$F(x,t) = \frac{1}{\sqrt{2\pi}} \int_{-\infty}^{+\infty} E_{\alpha,1}(-t^\alpha k^2) \, \tilde{f}(k) \, e^{i\,x\,k} dk, \quad (7.8.20)$$

where $\tilde{f}(k)$ is the Fourier transform[t] of $f(x)$ [155].

Proof. $\forall x \in \mathbb{R}, \forall \alpha \in \mathbb{R}^+, \forall t \in \mathbb{R}_0^+$, the action of the *PEO* $E_{\alpha,1}(t^\alpha \partial_x^2)$, on the initial function $f(x)$, is easily obtained by defining $f(x)$ through the Fourier transform and anti-transform,

$$\begin{aligned} \tilde{f}(k) &= \frac{1}{\sqrt{2\pi}} \int_{-\infty}^{\infty} f(x) e^{-ikx} dx, \\ f(x) &= \frac{1}{\sqrt{2\pi}} \int_{-\infty}^{\infty} \tilde{f}(k) e^{ixk} dk, \end{aligned} \quad (7.8.21)$$

therefore, using Eqs. (7.8.21) and the theorem of series integration, we have

$$E_{\alpha,1}(t^\alpha \, \partial_x^2) f(x) = \frac{1}{\sqrt{2\pi}} \int_{-\infty}^{\infty} E_{\alpha,1}(t^\alpha \, \partial_x^2) \tilde{f}(k) e^{ixk} dk. \quad (7.8.22)$$

Then, using *ML* definition (1.3.1), we obtain

$$\begin{aligned} F(x,t) &= E_{\alpha,1}(t^\alpha \partial_x^2) \, f(x) = \frac{1}{\sqrt{2\pi}} \int_{-\infty}^{\infty} E_{\alpha,1}(t^\alpha \, \partial_x^2) \tilde{f}(k) e^{ixk} dk \\ &= \frac{1}{\sqrt{2\pi}} \cdot \int_{-\infty}^{\infty} \sum_{r=0}^{\infty} \frac{t^{\alpha r} \partial_x^{2r}}{\Gamma(\alpha r + 1)} \tilde{f}(k) e^{ixk} dk \\ &= \frac{1}{\sqrt{2\pi}} \int_{-\infty}^{\infty} \sum_{r=0}^{\infty} \frac{t^{\alpha r}}{\Gamma(\alpha r + 1)} (ik)^{2r} \tilde{f}(k) e^{ixk} dk \\ &= \frac{1}{\sqrt{2\pi}} \int_{-\infty}^{\infty} \sum_{r=0}^{\infty} \frac{t^{\alpha r} (-k^2)^r}{\Gamma(\alpha r + 1)} \tilde{f}(k) e^{ixk} dk, \quad (7.8.23) \end{aligned}$$

---

[t]We observe that Eq. (7.8.20) can be recast in terms of Lévy distribution according to [100].

thus finally getting

$$F(x,t) = \frac{1}{\sqrt{2\pi}} \int_{-\infty}^{\infty} E_{\alpha,1}(-t^{\alpha}\,k^2)\tilde{f}(k)e^{ixk}dk. \qquad (7.8.24)$$

$\square$

Examples of solutions (7.8.20) are reported in Fig. 7.3, at different times for different α values, which clearly display a behavior which is not simply diffusive but also *anomalous*.

As already stressed, Eq. (7.8.18) is a fractional diffusive equation. This type of equations have played an increasingly important role in the description of processes called *super* or *sub-diffusive*, regarding the evolution of distributions whose mean square exhibits a dependence on time provided by a power law (faster than linear for the super diffusive case and vice versa for the sub-diffusive counterpart).

To better appreciate how these effects emerge from the previous formalism, we consider, $\forall x \in \mathbb{R}, \forall t \in \mathbb{R}_0^+$, the *ordinary heat diffusion equation*

$$\begin{cases} \partial_t F(x,t) = \partial_x^2 F(x,t) \\ F(x,0) = f(x), \end{cases} \qquad (7.8.25)$$

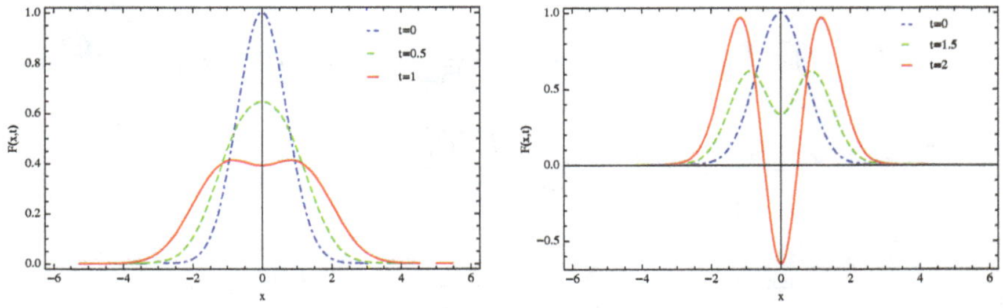

(a) $\alpha = 1.5$ (b) $\alpha = 3.5$

Figure 7.3: Solution $F(x,t)$ of Eq. (7.8.18) for $f(x) = e^{-x^2} \rightarrow \tilde{f}(k) = \dfrac{e^{-\frac{k^2}{4}}}{\sqrt{2}}$, at different times for different α values.

whose solution can be expressed in terms of Gauss–Weierstrass transform (1.2.2) (a direct consequence of the Fourier transform method, (7.8.21)), namely [18]

$$F(x,t) = \frac{1}{2\sqrt{\pi t}} \int_{-\infty}^{+\infty} e^{-\frac{(\xi-x)^2}{4t}} f(\xi)\, d\xi, \qquad (7.8.26)$$

where the distribution $f(x)$ is assumed to be normalized to unity with momenta

$$m_n(0) = \int_{-\infty}^{+\infty} x^n f(x)\, dx, \qquad \forall n \in \mathbb{N}. \qquad (7.8.27)$$

The moments associated with the distribution $F(x,t)$ are therefore specified by

$$m_n(t) = \int_{-\infty}^{+\infty} x^n F(x,t)\, dx = \int_{-\infty}^{+\infty} e^{-\frac{\xi^2}{4t}} f(\xi) I_n(\xi)\, d\xi,$$

$$I_n(\xi) = \frac{1}{2\sqrt{\pi}\, t} \int_{-\infty}^{+\infty} x^n e^{-\frac{x^2}{4t}} e^{\frac{x\xi}{2t}}\, dx.$$

$$(7.8.28)$$

The integral $I_n(\xi)$ can be evaluated using the generating function method [18], theorem of the series integration and the *GWI* (1.2.2), in fact

$$\sum_{n=0}^{\infty} \frac{u^n}{n!} I_n(\xi) = \frac{1}{2\sqrt{\pi}\, t} \int_{-\infty}^{+\infty} \sum_{n=0}^{\infty} \frac{(ux)^n}{n!} e^{-\frac{x^2}{4t}} e^{\frac{x\xi}{2t}}\, dx$$

$$= \frac{1}{2\sqrt{\pi}\, t} \int_{-\infty}^{+\infty} e^{\left(u+\frac{\xi}{2t}\right)x} e^{-\frac{x^2}{4t}}\, dx = e^{\frac{\xi^2}{4t}} e^{u\xi + u^2 t}$$

$$(7.8.29)$$

and, by the use of the generating function of two-variable Hermite polynomials (1.2.6), yields

$$I_n(\xi) = e^{\frac{\xi^2}{4t}} H_n(\xi, t), \qquad (7.8.30)$$

thus finally getting, by using Eq. (7.8.27),

$$m_n(t) = \int_{-\infty}^{+\infty} H_n(\xi, t) f(\xi) d\xi = H_n(\hat{m}, t)\mu_0,$$

$$H_n(\hat{m}, t) = n! \sum_{r=0}^{\lfloor \frac{n}{2} \rfloor} \frac{\hat{m}^{n-2r} t^r}{(n-2r)! r!}. \tag{7.8.31}$$

In Eq. (7.8.31), we have assumed that $\hat{m}$ is a kind of umbral operator acting on the vacuum μ_0 and defining the momenta as

$$\hat{m}^n \mu_0 = m_n(0). \tag{7.8.32}$$

In conclusion, we find

$$m_n(t) = n! \sum_{r=0}^{\lfloor \frac{n}{2} \rfloor} \frac{m_{n-2r}(0) t^r}{(n-2r)! r!}, \tag{7.8.33}$$

where $m_2(t)$ shows a linear dependence on time.

The formalism we have developed so far yields the possibility of evaluating the momenta associated with the distribution (7.8.19)–(7.8.20) by the use of the following substitution:

$$H_n(\hat{m}, t) \rightarrow H_n\left(\hat{m}, {}_{\alpha,1}\hat{d}\, t^\alpha\right). \tag{7.8.34}$$

The second momentum is, in this case, nonlinear and, recalling Eqs. (1.3.13)–(1.3.14), we introduce the following polynomials whose properties will be discussed in Section 7.8.1.

Definition 35. $\forall x, y \in \mathbb{R}, \forall \alpha \in \mathbb{R}^+$, the family of polynomials

$$_\alpha H_n(x, y) := H_n\left(x, {}_{\alpha,1}\hat{d}\, y\right)\psi_0 = n! \sum_{r=0}^{\lfloor \frac{n}{2} \rfloor} \frac{x^{n-2r} \left({}_{\alpha,1}\hat{d}\, y\right)^r}{(n-2r)! r!}\psi_0$$

$$= n! \sum_{r=0}^{\lfloor \frac{n}{2} \rfloor} \frac{x^{n-2r} y^r}{(n-2r)! \Gamma(\alpha r + 1)}, \tag{7.8.35}$$

is called *Mittag–Leffler–Hermite (MLH)*.

The introduction of the *PEO*, $E_{\alpha,1}\left(t^\alpha \partial_x^2\right)$, is of central importance for our forthcoming discussion, its role and underlying computational rules will be therefore carefully explored in the following.

In order to provide further elements allowing to appreciate the flexibility of the procedure employing the umbral methods, we consider the *fractional Poisson distribution (FPD)*, discussed in [66], within the context of non-Markovian stochastic processes with a non-exponential distribution of inter-arrival times.

90 Without entering into the Theory of the fractional Poisson processes, we note that the equation governing the generating function of the distribution itself is given, $\forall \alpha \in \mathbb{R}^+, \forall s \in \mathbb{R}, \forall t \in \mathbb{R}_0^+$, by [117]

$$G_\alpha(s,t) = E_{\alpha,1}\left(-(1-s)\,\Omega\,t^\alpha\right), \qquad (7.8.36)$$

Ω has physical dimension $[\Omega] = \frac{1}{T^\alpha}$, where T is the time. By the use of the umbral notation (1.3.14), we can expand the previous generating function as

$$G_\alpha(s,t) = \sum_{m=0}^{\infty} s^m\,_\alpha P(m,t), \qquad (7.8.37)$$

where

$$_\alpha P(m,t) = \frac{(\Omega t^\alpha)^m}{m!} \sum_{n=0}^{\infty} \frac{(n+m)!}{\Gamma(\alpha(n+m)+1)} \frac{(-\Omega t^\alpha)^n}{n!} \qquad (7.8.38)$$

is the *FPD*, introduced in [167]. With this aim, we use the framework of the umbral formalism and Eqs. (7.8.36)–(1.3.14)–(1.1.7)–exponential series expansion–(1.3.13).

$$G_\alpha(s,t) = E_{\alpha,1}\left(-(1-s)\,\Omega\,t^\alpha\right) = e^{\alpha,1\hat{d}\left(-(1-s)\Omega t^\alpha\right)}\psi_0$$

$$= e^{\,\alpha,1\hat{d}\,s\,(\Omega\,t^\alpha)}e^{-\,\alpha,1\hat{d}\,(\Omega\,t^\alpha)}\psi_0$$

$$= \sum_{m=0}^{\infty} s^m \frac{_\alpha\hat{d}^{\,m}}{m!}(\Omega t^\alpha)^m \sum_{n=0}^{\infty} \frac{_\alpha\hat{d}^{\,n}}{n!}(-\Omega t^\alpha)^n\psi_0$$

$$= \sum_{m=0}^{\infty} s^m \frac{(\Omega t^\alpha)^m}{m!} \sum_{n=0}^{\infty} \frac{(-\Omega t^\alpha)^n}{n!} {}_{\alpha,1}\hat{d}^{m+n}\psi_0$$

$$= \sum_{m=0}^{\infty} s^m \frac{(\Omega t^\alpha)^m}{m!} \sum_{n=0}^{\infty} \frac{(-\Omega t^\alpha)^n}{n!} \frac{(n+m)!}{\Gamma(\alpha(n+m)+1)}$$

$$= \sum_{m=0}^{\infty} s^m {}_\alpha P(m,t), \tag{7.8.39}$$

91 According to the methods we have envisaged, to calculate average and r.m.s. values, we proceed as follows. We set, $\forall \alpha, \Omega \in \mathbb{R}^+, \forall t \in \mathbb{R}_0^+, \forall m \in \mathbb{N}$,

$$_\alpha P(m,t) = \frac{\left({}_{\alpha,1}\hat{d}\,\Omega\,t^\alpha\right)^m}{m!} e^{-{}_{\alpha,1}\hat{d}\,\Omega\,t^\alpha}\psi_0, \tag{7.8.40}$$

so finding, for the first-order moment,

$$\text{(i)} \ \langle\, {}_\alpha m_t \,\rangle = \frac{(\Omega\,t^\alpha)}{\Gamma(\alpha+1)} \tag{7.8.41}$$

and, for the variance,

$$\text{(ii)} \ {}_\alpha\sigma_t^2 = \frac{2\,(\Omega\,t^\alpha)^2}{\Gamma(2\alpha+1)} + \frac{(\Omega\,t^\alpha)}{\Gamma(\alpha+1)} - \frac{(\Omega\,t^\alpha)^2}{(\Gamma(\alpha+1))^2}, \tag{7.8.42}$$

in agreement with the results obtained in [117, 167]. Indeed, $\forall \alpha, \Omega \in \mathbb{R}^+, \forall t \in \mathbb{R}_0^+, \forall m \in \mathbb{N}$, by using Eqs. (1.3.14)–(1.3.13) and algebraic manipulation, we obtain

$$\text{(i)} \ \langle\, {}_\alpha m_t \,\rangle = \sum_{m=0}^{\infty} m \frac{({}_\alpha\hat{d}\,\Omega\,t^\alpha)^m}{m!} e^{-{}_\alpha\hat{d}\,\Omega\,t^\alpha}\psi_0$$

$$= \sum_{m=1}^{\infty} ({}_\alpha\hat{d}\,\Omega\,t^\alpha) \frac{({}_\alpha\hat{d}\,\Omega\,t^\alpha)^{m-1}}{(m-1)!} e^{-{}_\alpha\hat{d}\,\Omega\,t^\alpha}\psi_0$$

$$= ({}_\alpha\hat{d}\,\Omega\,t^\alpha) e^{{}_\alpha\hat{d}\,\Omega\,t^\alpha} e^{-{}_\alpha\hat{d}\,\Omega\,t^\alpha}\psi_0$$

$$= e^{{}_\alpha\hat{d}\,\Omega\,t^\alpha}\psi_0 = \frac{(\Omega\,t^\alpha)}{\Gamma(\alpha+1)}. \tag{7.8.43}$$

$$(ii) \ _\alpha\sigma_t^2 = \sum_{m=0}^{\infty} m^2 \, _\alpha P(m,t) - \left(\sum_{m=0}^{\infty} m \, _\alpha P(m,t) \right)^2$$

$$= \sum_{m=1}^{\infty} m \frac{(_\alpha\hat{d} \, \Omega \, t^\alpha)^m}{(m-1)!} e^{-_\alpha\hat{d} \, \Omega \, t^\alpha} \psi_0 - \left(\frac{(\Omega \, t^\alpha)}{\Gamma(\alpha+1)} \right)^2$$

$$= \sum_{m=1}^{\infty} \frac{(m-1+1)(_\alpha\hat{d} \, \Omega \, t^\alpha)^m}{(m-1)!} e^{-_\alpha\hat{d} \, \Omega \, t^\alpha} \psi_0 - \left(\frac{(\Omega \, t^\alpha)}{\Gamma(\alpha+1)} \right)^2$$

$$= \left[(_\alpha\hat{d} \, \Omega \, t^\alpha)^2 \sum_{m=2}^{\infty} \frac{(_\alpha\hat{d} \, \Omega \, t^\alpha)^{m-2}}{(m-2)!} \right.$$

$$\left. + (_\alpha\hat{d} \, \Omega \, t^\alpha) \sum_{m=1}^{\infty} \frac{(_\alpha\hat{d} \, \Omega \, t^\alpha)^{m-1}}{(m-1)!} \right]$$

$$]3pt[\quad \cdot e^{-_\alpha\hat{d} \, \Omega \, t^\alpha} \psi_0 - \left(\frac{(\Omega \, t^\alpha)}{\Gamma(\alpha+1)} \right)^2$$

$$= (_\alpha\hat{d} \, \Omega \, t^\alpha)^2 \psi_0 + (_\alpha\hat{d} \, \Omega \, t^\alpha) \psi_0 - \left(\frac{(\Omega \, t^\alpha)}{\Gamma(\alpha+1)} \right)^2$$

$$= \frac{2(\Omega \, t^\alpha)^2}{\Gamma(2\alpha+1)} + \frac{(\Omega \, t^\alpha)}{\Gamma(\alpha+1)} - \frac{(\Omega \, t^\alpha)^2}{(\Gamma(\alpha+1))^2}. \tag{7.8.44}$$

7.8.1 *Mittag–Leffler–Hermite Polynomials*

In Eq. (7.8.35), we have introduced a family of polynomials that we have called *MLH*. They play, within the context of *fractional diffusion heat equation* (*FDHE*), the same role of the heat polynomials [178] in the case of the ordinary heat equation.

Proposition 52. *Let us consider the fractional evolution problem* (7.8.18) *with initial condition* $F(x,0) = x^n, \forall n \in \mathbb{N}$, *the relevant solution can accordingly be written as*

$$_\alpha H_n(x,t^\alpha) = \left(e^{\, _{\alpha,1}\hat{d} \, t^\alpha \partial_x^2} x^n \right) \psi_0 \, . \tag{7.8.45}$$

Proof. By the use of Eq. (7.8.35), we can write $\forall x \in \mathbb{R}, \forall \alpha \in \mathbb{R}^+, \forall t \in \mathbb{R}_0^+, \forall n \in \mathbb{N}$,

$$_\alpha H_n(x, t^\alpha) = H_n\left(x, {}_{\alpha,1}\hat{d}\, t^\alpha\right)\psi_0 = n! \sum_{r=0}^{\lfloor \frac{n}{2} \rfloor} \frac{x^{n-2r} t^{\alpha r}}{(n-2r)!\Gamma(\alpha r + 1)}$$

$$(7.8.46)$$

but, on the other hand, expanding the exponential, acting on the successive derivatives of x^n and applying the umbral operator (1.3.13), we find

$$\left(e^{\,{}_{\alpha,1}\hat{d}\, t^\alpha \partial_x^2}\, x^n\right)\psi_0 = \sum_{r=0}^{\infty} \frac{{}_{\alpha,1}\hat{d}^{\,r}\, t^{\alpha r}\, \partial_x^{2r}}{r!}\, x^n\,\psi_0$$

$$= \sum_{r=0}^{\lfloor \frac{n}{2} \rfloor} \frac{n!\, x^{n-2r} t^{\alpha r}\, {}_\alpha \hat{d}^{\,r}}{(n-2r)!\, r!}\psi_0 = n! \sum_{r=0}^{\lfloor \frac{n}{2} \rfloor} \frac{x^{n-2r} t^{\alpha\, r}}{(n-2r)!\Gamma(\alpha\, r + 1)}$$

$$(7.8.47)$$

$\square$

The *MLH* belong to the *Appéll polynomial* family [5] and are easily shown to satisfy, in its domain, the recurrences (see [18] where the relevant properties have been touched upon on and proved.)

Properties 25.

(i) $\partial_x\, {}_\alpha H_n(x, t^\alpha) = n\, {}_\alpha H_{n-1}(x, t^\alpha);$ (7.8.48)

(ii) $\partial_t\, {}_\alpha H_n(x, t^\alpha) = n(n-1){}_\alpha H_{n-2}(x, t^\alpha) + \dfrac{t^{-\alpha}}{\Gamma(1-\alpha)} x^n$.   (7.8.49)

The generating function can be obtained from Eq. (7.8.45) as follows.

Corollary 50. *By Proposition 52, exploiting Eqs.* (1.2.6)–(1.3.13)– (1.3.14),

$$\sum_{n=0}^{\infty} \frac{\xi^n}{n!}{}_\alpha H_n(x, t^\alpha) = \sum_{n=0}^{\infty} \frac{\xi^n}{n!} H_n\left(x, {}_{\alpha,1}\hat{d}\, t^\alpha\right)\psi_0 = \left(e^{x\,\xi}\, e^{\,{}_{\alpha,1}\hat{d}\, t^\alpha \xi^2}\right)\psi_0$$

$$= E_{\alpha,1}\left(t^\alpha \xi^2\right) e^{x\,\xi}.$$

$$(7.8.50)$$

In order to write Eq. (7.8.50) in a more convenient form for our purposes, we note the following identity [21, 100]:

$$\int_0^\infty n_\alpha(s,t)\frac{s^n}{n!}ds = \frac{t^{\alpha\,n}}{\Gamma(\alpha\,n+1)}, \quad \forall n \in \mathbb{N}, \forall \alpha \in \mathbb{R}^+, \forall t \in \mathbb{R}_0^+,$$

$$n_\alpha(s,t) := \frac{1}{\alpha}\frac{1}{s\sqrt[\alpha]{s}}\,g_\alpha\left(\frac{t}{\sqrt[\alpha]{s}}\right).$$

$$(7.8.51)$$

with $g_\alpha(x)$ being the one-sided *Lévy* stable distribution.

92 From the previous identities we may draw at least two conclusions (see [100]):

(i) $_\alpha H_n(x,t^\alpha) = \int_0^\infty n_\alpha(s,t)H_n(x,s)ds,$ $(7.8.52)$

(ii) $E_{\alpha,1}(b\,t^\alpha) = \int_0^\infty n_\alpha(s,t)e^{b\,s}ds.$ $(7.8.53)$

Equation (7.8.53) can be exploited to cast the *PEO* associated to the fractional evolution problem (7.8.18) in the form

$$E_{\alpha,1}\left(t^\alpha\partial_x^2\right) = \int_0^\infty n_\alpha(s,t)e^{s\,\partial_x^2}ds.$$

$$(7.8.54)$$

Since $e^{s\,\partial_x^2}$ is the evolution operator for the ordinary diffusion problem, we can write the solution (7.8.19) of problem (7.8.18) as

$$F(x,t) = \int_0^\infty n_\alpha(s,t)\left(e^{s\partial_x^2}f(x)\right)ds$$

$$(7.8.55)$$

and, in the case in which $f(x) = e^{-x^2}$, by the use of Eq. (1.2.2) we obtain the *Glaisher* identity [67]

$$e^{s\partial_x^2}e^{-x^2} = \frac{1}{\sqrt{1+4s}}e^{-\frac{x^2}{1+4s}}, \quad \forall x, s \in \mathbb{R},$$

$$(7.8.56)$$

which yields

$$F(x,t) = \int_0^\infty \frac{n_\alpha(s,t)}{\sqrt{1+4s}}e^{-\frac{x^2}{1+4s}}ds.$$

$$(7.8.57)$$

93 A further important conclusion which may be drawn from the previous formalism concerns the question whether the family of polynomials *MLH* (7.8.35) can be considered orthogonal. The question can be settled out in a fairly simple way by assuming that an expansion of the type

$$G(x, \xi) = \sum_{n=0}^{\infty} a_n \, _\alpha H_n \left(x, \xi^\alpha \right), \quad \forall x, \xi \in \mathbb{R}, \forall \alpha \in \mathbb{R}^+. \quad (7.8.58)$$

be allowed. According to Eq. (7.8.52), we find

$$G(x, \xi) = \sum_{n=0}^{\infty} a_n \int_0^\infty n_\alpha(s, \xi) H_n(x, s) ds. \quad (7.8.59)$$

Thanks to the previous conclusions, we can therefore state the following theorem.

Theorem 7. *If $\forall x, s \in \mathbb{R}$, the series $\sum_{n=0}^{\infty} a_n \, H_n(x, s)$ is uniformly converging to a function $g(x, s)$, then, by MLH-definition, the expansion (7.8.58) does exist and the expansion coefficients are the same as for the ordinary expansion.*

7.9   Voigt Functions

In the previous sections, we have provided some hints on the use of the Hermite calculus to study integral forms with specific application in different fields of research. In this section, we will show how the method can be extended to a systematic investigation of the Voigt functions and to their relevant generalizations [138].

The Hermite and Laguerre functions are the extension of the corresponding polynomials to negative and/or real indices. In this paragraph, we apply the proposed method for the study of the *Voigt functions* which find several applications in spectroscopy [165] and are defined by the convolution of Gaussian and Lorentzian distributions. Apart from their interest in Physics, they have raised a certain interest in Mathematics for their relationship with a number of special functions.

94 We denote the *Voigt funtions* (VF) by $K(x,y,z)$, $L(x,y,z)$ and define them in terms of the integral representations [136] $\forall x, z \in \mathbb{R}, \forall y \in \mathbb{R}^+$

$$K(x,y,z) = \frac{1}{\sqrt{\pi}} \int_0^\infty e^{-x\xi - y\xi^2} \cos(z\xi) \, d\xi ,$$

$$L(x,y,z) = \frac{1}{\sqrt{\pi}} \int_0^\infty e^{-x\xi - y\xi^2} \sin(z\xi) \, d\xi \qquad (7.9.1)$$

(the definition in [136] includes two variables only (x, z) and assumes $y = \frac{1}{4}$). If we introduce the complex VF

$$E(x,y,z) = \frac{1}{\sqrt{\pi}} \int_0^\infty e^{-(x-iz)\xi - y\xi^2} \, d\xi \qquad (7.9.2)$$

and define VF in Eq. (7.9.1) as the relevant real and imaginary parts, we can easily conclude that it is expressible in terms of "erfc" function. We can indeed exploit Eqs. (2.3.17)–(2.3.18) to end up with the identity

$$E(x,y,z) = \frac{1}{\sqrt{\pi}} \int_0^\infty e^{-(x-iz)\xi - y\xi^2} \, d\xi = \frac{1}{\sqrt{\pi}} H_{-1}(x - iz, y)$$

$$= \frac{1}{\sqrt{\pi}} \left(x - iz + y\hat{h} \right)^{-1} \theta_0$$

$$= \frac{1}{2\sqrt{y}} e^{\frac{(x-iz)^2}{4y}} erfc\left(\frac{x - iz}{2\sqrt{y}} \right). \qquad (7.9.3)$$

95 We obtain the relevant derivatives by the use of the well-known properties of the Hermite functions, $\forall m \in \mathbb{N}$

$$\partial_z^m E(x,y,z) = \frac{i^m \, m!}{\sqrt{\pi}} H_{-(m+1)}(x - iz, y). \qquad (7.9.4)$$

We state the following definition to develop Exercise 96.

Definition 36. We introduce the *Voigt (V-) transform* of a function $f(z)$

$$_V\hat{f}(x,y;z) = \int_0^\infty e^{-xt - yt^2} f(zt) \, dt, \qquad (7.9.5)$$

which can be viewed as a generalized form of transform.

The procedure we have envisaged allows to unify many of the previous analyses [139] aimed at getting a different way of expressing the *VF* in forms suitable for various specific applications.

96 By using the introduced Voigt transform, $\forall x, y, z \in \mathbb{R}, \forall n \in \mathbb{N}$, we get

$$_V\hat{f}(x, y; z) = \sum_{n=0}^{\infty} a_n H_{-(n+1)}(x, -y) z^n, \qquad f(z) = \sum_{n=0}^{\infty} a_n \frac{z^n}{n!}.$$
$$(7.9.6)$$

If we use the identity (1.1.16) $t^{z\partial_z} f(z) = f(tz)$, Laplace transform and the technique of Example 17, we can write

$$_V\hat{f}(x, y; z) = \int_0^{\infty} e^{-xt - yt^2} t^{z\partial_z} \, dt \, f(z)$$

$$= \Gamma(z\partial_z + 1) H_{-(z\partial_z + 1)}(x, -y) f(z). \quad (7.9.7)$$

By expanding the function $f(z)$ in series and by using the property [18] $f(z\partial_z) z^n = f(n) z^n$, we can finally write Eq. (7.9.7).

97 According to the present formalism, the *VF* (7.9.1) is the *V*-transform (7.9.5) of the *circular functions*, thus reading

$$K(x, y, z) = \sum_{n=0}^{\infty} (-1)^n H_{-(2n+1)}(x, -y) z^{2n},$$
$$(7.9.8)$$
$$L(x, y, z) = z \sum_{n=0}^{\infty} (-1)^n H_{-(2n+2)}(x, -y) z^{2n}.$$

98 *Let us furthermore note that the V-transform of a zeroth-order cylindrical Bessel function is*

$$_V\hat{f}(x, y; z) = \sum_{n=0}^{\infty} \frac{(2n)!}{n!^2} (-1)^n H_{-(2n+1)}(x, -y) \left(\frac{z}{2}\right)^{2n}.$$
$$(7.9.9)$$

The generalization of the V-functions proposed in [158] can be viewed as V-transforms of different families of functions.

99 In Chapter 2, we have noted that Hermite functions of negative order can be defined by means of infinite integrals yielding the relevant integral representation, however, the use of the formalism we are proposing may be useful in a wider context as, e.g., for evaluation of definite integrals as follows:

$$\int_0^x \xi^{\nu-1} e^{-a\xi - b\xi^2} d\xi = \int_0^x \xi^{\nu-1} e^{-(a + _{-|b|}\hat{h})\xi} d\xi \, \theta_0$$

$$= \frac{1}{\left(a + _{-|b|}\hat{h}\right)^\nu} \gamma\left(\nu, \left(a + _{-|b|}\hat{h}\right)x\right) \theta_0$$

$$= \sum_{n=0}^\infty \frac{(-1)^n H_n(a, -b) x^{\nu+n}}{n!(\nu + n)}, \qquad (7.9.10)$$

with $\gamma(\nu, x)$ being the *incomplete gamma function* (7.1.50) [3]. Let us therefore consider the following V-transform:

$$_V\hat{f}_\mu(x, y, z) = \int_0^\infty t^{\mu-1} e^{-xt - yt^2} J_0(zt) \, dt, \qquad (7.9.11)$$

which, according to the previous discussion, can formally be written as

$$_V\hat{f}_\mu(x, y, z) = \int_0^\infty t^{\mu-1+z\partial_z} e^{-xt - yt^2} \, dt \, J_0(z)$$

$$= \sum_{n=0}^\infty \frac{\Gamma(2n + \mu)}{n!^2} (-1)^n H_{-(2n+\mu)}(x, -y) \left(\frac{z}{2}\right)^{2n}$$

$$= e^{-\hat{g}\left(\frac{z}{2}\right)^2} \Psi_0,$$

$$\hat{g}^n \, \Psi_0 = \frac{\Gamma(2n + \mu)}{n!} H_{-(2n+\mu)}(x, -y).$$

$$(7.9.12)$$

This umbral form can be exploited, e.g., to derive the integral

$$\int_{-\infty}^\infty {}_V\hat{f}_\mu(x,y,z) dz = 2\sqrt{\pi}\hat{g}^{-\frac{1}{2}} \, \Psi_0 = 2\sqrt{\pi} \frac{\Gamma(\mu-1)}{\Gamma\left(\frac{1}{2}\right)} H_{-(\mu-1)}(x,-y)$$

$$(7.9.13)$$

as also checked by a direct integration of Eq. (7.9.11), namely

$$\int_{-\infty}^\infty {}_V\hat{f}_\mu(x, y, z) \, dz = \int_0^\infty t^{\mu-1} e^{-xt - yt^2} \left(\int_{-\infty}^\infty J_0(zt) \, dz\right) dt$$

$$= 2 \int_0^\infty t^{\mu-2} e^{-xt - yt^2} \, dt. \qquad (7.9.14)$$

7.10   Applications to Physics

The matters we discussed so far deal only with Mathematics. No applicative problems have been touched upon, except the fact that many specific issues, like the evaluation of complex integrals, have been greatly simplified.

In this section, we treat some applications of the formalism to physical problems as, e.g., Classical Optics, Laser Physics and Relativistic Quantum Mechanics.

We start with an example regarding the fractional form of the Schrödinger equation (*FSE*). Albeit fractional evolution equations are getting significant attention for the study of physical phenomena like anomalous diffusion, the case of the Schrödinger equation is still at the level of mere speculation.

7.10.1   *Fractional Schrödinger Equation*

In Section 7.3, we introduced the fractional Poisson distribution (*FPD*) (7.8.38) as a mere consequence of the definition of the *ML* function. In the following, we derive a different form of *FPD* by solving the *FSE* for a hypothetical process implying the emission and absorption of photons.

101 We assume that the relevant dynamics is ruled by the *ML* - Schrödinger equation[u] [129]

$$i^\alpha \, \partial_t^\alpha \mid \Psi \, \rangle = \hat{H} \mid \Psi \, \rangle + i^\alpha \frac{t^{-\alpha}}{\Gamma(1-\alpha)} \mid \Psi(0)\rangle,$$

$$\hat{H} = i^\alpha \, \Omega \left(\hat{a} - \hat{a}^+ \right), \qquad 0 \le \alpha \le 1, \quad \forall t \in \mathbb{R}_0^+$$

(7.10.1)

where $\hat{a}, \hat{a}^+$ are annihilation, creation operators, respectively, [119] satisfying the commutation relation $[\hat{a}, \hat{a}^+] = \hat{1}$ and the

---

[u]According to Dirac notation, we write the state $\mid \Psi \rangle$ to indicate the function $\Psi(t)$.

constant Ω in Eq. (7.10.1) has the dimension of $t^{-\alpha}$. If we work on a *Fock* basis [119] and choose the "physical" vacuum (namely the state of the quantized electromagnetic field with no photons) as the initial state of our process (7.10.1), namely $| \Psi(t) \rangle |_{t=0} = | 0 \rangle$, we can understand how the field ruled by an *FSE* evolves from the vacuum. The comparison with the ordinary Schrödinger counterpart is interesting, because the field evolves into a coherent state, displaying an emission process in which the photon counting statistics follows a Poisson distribution [119]. The formal solution of the evolution problem provided by Eq. (7.10.1) is formally obtained, using[v] Eqs. (1.3.14)–(7.8.19) as[w] [66]

$$| \Psi \rangle = e^{\,_\alpha \hat{d}\, t^\alpha\, \Omega\, (\hat{a} - \hat{a}^+)} | 0 \rangle, \qquad (7.10.2)$$

in which the evolution operator

$$\hat{U}(t) = e^{\,_\alpha \hat{d}\, \Omega\, t^\alpha (\hat{a} - \hat{a}^+)} \qquad (7.10.3)$$

cannot be straightforwardly disentangled (namely written as product of exponential operators) since the creation and annihilation operators are not commuting.[x] According to the previous prescription, we set

$$_\alpha \hat{A} = {}_\alpha \hat{d}\, \Omega\, t^\alpha \hat{a}, \qquad e^{{}_\alpha \hat{d}\, \Omega\, t^\alpha (\hat{a} - \hat{a}^+)} = e^{\,_\alpha \hat{A} - {}_\alpha \hat{A}^+}. \qquad (7.10.4)$$

We note that

$$\left[{}_\alpha \hat{A},\ {}_\alpha \hat{A}^+ \right] = {}_\alpha \hat{d}^{\,2} \left(\Omega\, t^\alpha \right)^2. \qquad (7.10.5)$$

---

[v] Here, we omit the ${}_\alpha \hat{d}$-vacuum ψ_0 (1.3.12).

[w] To simplify the writing we substitute ${}_{\alpha,1} \hat{d}$ with ${}_\alpha \hat{d}$.

[x] With this aim, we recall the Weyl [57] disentanglement rules $e^{\hat{A} + \hat{B}} = e^{\hat{A}} e^{\hat{B}} e^{-\frac{\hat{k}}{2}}$, $\left[\hat{A}, \hat{B} \right] = \hat{A}\hat{B} - \hat{B}\hat{A} = \hat{k}$, $\left[\hat{A}, \hat{k} \right] = \left[\hat{k}, \hat{B} \right] = 0$, which hold if $\hat{A}, \hat{B}$ are not commuting with each other, but their commutator "commutes" both with $\hat{A}$ and $\hat{B}$.

By the use of Weyl and algebraic rules, we get

$$\left[{}_\alpha\hat{A}, \, {}_\alpha\hat{A}^+\right] = {}_\alpha\hat{A}{}_\alpha\hat{A}^+ - {}_\alpha\hat{A}^+_\alpha\hat{A}$$

$$= {}_\alpha\hat{d}\,\Omega\,t^\alpha\hat{a}\,{}_\alpha\hat{d}\,\Omega\,t^\alpha\hat{a}^+ - {}_\alpha\hat{d}\,\Omega\,t^\alpha\hat{a}^+\,{}_\alpha\hat{d}\,\Omega\,t^\alpha\hat{a}$$

$$= \left({}_\alpha\hat{d}\,\Omega\,t^\alpha\right)^2 \left(\hat{a}\,\hat{a}^+ - \hat{a}^+\hat{a}\right)$$

$$= \left({}_\alpha\hat{d}\,\Omega\,t^\alpha\right)^2 \left[\hat{a}, \hat{a}^+\right] = {}_\alpha\hat{d}^2\,(\Omega\,t^\alpha)^2. \qquad (7.10.6)$$

Since the umbral operator ${}_\alpha\hat{d}$ is independent of the creation annihilation counterparts, we end up with the exponential disentanglement[y]

$$e^{{}_\alpha\hat{A}-{}_\alpha\hat{A}^+} = e^{-\frac{(\Omega\,t^\alpha)^2\,{}_\alpha\hat{d}^2}{2}}\,e^{-{}_\alpha\hat{A}^+}e^{{}_\alpha\hat{A}}. \qquad (7.10.7)$$

The solution of our *FSE* therefore can be written as [66]

$$|\,\Psi\,\rangle = e^{{}_\alpha\hat{d}\,t^\alpha\,\Omega\,\left(\hat{a}-\hat{a}^+\right)}\,|\,0\,\rangle$$

$$= e^{-\frac{\left({}_\alpha\hat{d}\,t^\alpha\,\Omega\right)^2}{2}}e^{-\left({}_\alpha\hat{d}\,t^\alpha\,\Omega\right)\hat{a}^+}e^{\left({}_\alpha\hat{d}\,t^\alpha\,\Omega\right)\hat{a}}\,|\,0\,\rangle.$$

$$(7.10.8)$$

The use of the identities [119]

$$(\hat{a}^+)^n\,|\,0\,\rangle = \sqrt{n!}\,|\,n\,\rangle, \qquad\qquad \hat{a}\,|\,0\,\rangle = 0, \qquad (7.10.9)$$

finally yields the solution in the form [66]

$$|\,\Psi\,\rangle = e^{-\frac{\left({}_\alpha\hat{d}\,t^\alpha\,\Omega\right)^2}{2}}e^{-\left({}_\alpha\hat{d}\,t^\alpha\,\Omega\right)\hat{a}^+}\,|\,0\,\rangle$$

$$= e^{-\frac{\left({}_\alpha\hat{d}\,t^\alpha\,\Omega\right)^2}{2}}\sum_{n=0}^{\infty}\frac{\left(-{}_\alpha\hat{d}\,t^\alpha\Omega\right)^n}{\sqrt{n!}}\,|\,n\,\rangle. \quad (7.10.10)$$

---

[y]We recall that the Weyl identity provides

$$\left[\hat{A}, \hat{B}\right] = \hat{k} \Rightarrow \left[\hat{B}, \hat{A}\right] = -\hat{k}; \qquad e^{\hat{A}+\hat{B}} = \begin{cases} e^{\hat{A}}e^{\hat{B}}e^{-\frac{\hat{k}}{2}}, \\ e^{\hat{B}}e^{\hat{A}}e^{\frac{\hat{k}}{2}}. \end{cases}$$

102 The use of the orthogonality properties of the number photon states $|m\rangle$, namely $\langle n \mid m \rangle = \delta_{n,m}$, yields probability amplitude of finding the state $\mid \Psi \rangle$ in a photon number state $|m\rangle$. It is just according to the identity [66]

$$\langle m \mid \Psi \rangle = e^{-\frac{(_\alpha \hat{d}\, t^\alpha\, \Omega)^2}{2}} \cdot \frac{\left(-_\alpha \hat{d}\, t^\alpha\, \Omega\right)^m}{\sqrt{m!}}, \qquad (7.10.11)$$

which is formally equivalent to a Poisson probability amplitude (7.8.40).

103 The probability distribution, $\forall X, m, \alpha \in \mathbb{R}_0^+ : \alpha \leq 1$ and $X = (t^\alpha\, \Omega)^2$, is then found as [66]

$$\begin{aligned}
_\alpha p(m,t) &= |\langle m \mid \Psi \rangle|^2 = e^{-\left(_\alpha \hat{d}\, t^\alpha\, \Omega\right)^2} \cdot \frac{\left(_\alpha \hat{d}\, t^\alpha\, \Omega\right)^{2m}}{m!} \\
&= \frac{X^m}{m!} e_m^{(\alpha,\, 2)}(-X), \\
e_m^{(\alpha,\, 2)}(-X) &= \sum_{r=0}^{\infty} \frac{(-1)^r}{r!} \frac{\Gamma(2(r+m)+1)}{\Gamma(2(r+m)\alpha + 1)} X^r, \qquad (7.10.12)
\end{aligned}$$

which is similar, but not equivalent, to the *FPD* derived in (7.8.38).

$$\begin{aligned}
_\alpha p(m,t) &= e^{-\left(_\alpha \hat{d}\, t^\alpha\, \Omega\right)^2} \cdot \frac{(_\alpha \hat{d}\, t^\alpha\, \Omega)^{2m}}{m!} \psi_0 \\
&= \sum_{r=0}^{\infty} \frac{(-1)^r}{r!} \frac{(t^\alpha \Omega)^{2r+2m}}{m!} \,_\alpha \hat{d}^{\,2r+2m} \psi_0 \\
&= \sum_{r=0}^{\infty} \frac{(-1)^r}{r!} \frac{(t^\alpha \Omega)^{2(r+m)}}{m!} \frac{\Gamma(2(r+m)+1)}{\Gamma(2(r+m)\alpha + 1)}. \qquad (7.10.13)
\end{aligned}$$

It is however evident that the probability (7.10.12) is properly normalized and indeed $\sum_{m=0}^{\infty} {}_\alpha p(m) = 1$.

104 Regarding the evaluation of the average number of emitted photons, we proceed as in the proof of (7.8.41).

$$\langle m \rangle = \sum_{m=0}^{\infty} m \, {}_\alpha p(m) = \sum_{m=0}^{\infty} m \frac{X^m}{m!} e_m^{(\alpha,\,2)}(-X)$$

$$= \sum_{m=1}^{\infty} \frac{X^m}{(m-1)!} \sum_{r=0}^{\infty} \frac{(-1)^r}{r!} {}_\alpha \hat{d}^{2(r+m)} X^r \, \psi_0$$

$$= e^{-\left({}_\alpha \hat{d}^2 X\right)} \sum_{m=1}^{\infty} \frac{\left({}_\alpha \hat{d}^2 X\right)^m}{(m-1)!} \psi_0 = {}_\alpha \hat{d}^2 X \, \psi_0 = \frac{2X}{\Gamma(2\alpha+1)},$$

$$(7.10.14)$$

and the analogous procedure (7.8.42) allows the evaluation of the r.m.s. of the emitted photons, namely

$$\sigma_m^2 = \langle\, m^2\, \rangle - \langle\, m\, \rangle^2 = {}_\alpha \hat{d}^4 X^2 + {}_\alpha \hat{d}^2 X - \left(\frac{2X}{\Gamma(2\alpha+1)}\right)^2$$

$$= 2X \left[2X \left(\frac{6}{\Gamma(4\alpha+1)} - \frac{1}{(\Gamma(2\alpha+1))^2} \right) + \frac{1}{\Gamma(2\alpha+1)} \right].$$

$$(7.10.15)$$

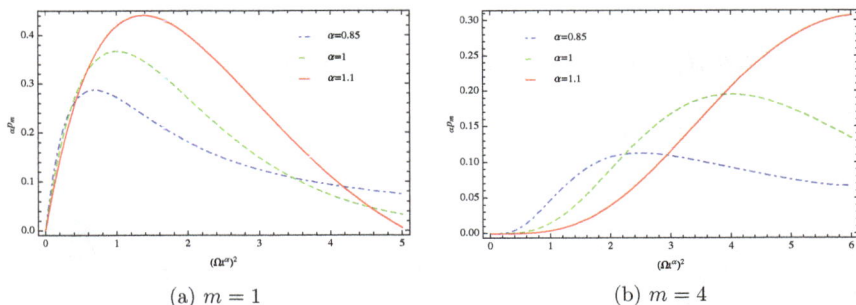

(a) $m = 1$ (b) $m = 4$

Figure 7.4: Probability distribution "${}_\alpha p(m)$" vs $(\Omega t^\alpha)^2$, for different values of α and m.

7.10.2 *Pearcey Integral in Optics and Evolutive Problems*

105 According to the formalism, of Section 7.2.1, the Pearcey
integral, studied in Optics, is easily reduced to a particular
case of Eq. (7.2.29), within the framework of diffraction
problems [120] so, $\forall x, y \in \mathbb{R}$,

$$J(1, x, -iy) = \int_{-\infty}^{\infty} e^{-(t^4 + xt^2) + iyt} dt = \sqrt{\frac{\pi}{\hat{h}_{(x,-1)}}} e^{-\frac{y^2}{4\hat{h}_{(x,-1)}}} \eta_0$$

$$(7.10.16)$$

and can be expressed in terms of parabolic cylinder functions,
as indicated in Exercise 50. It is perhaps worth stressing that,
in the literature, a converging series for the Pearcey integral
is given in the form [26]

$$J(1, x, -iy) = \int_{-\infty}^{\infty} e^{-t^4 - \hat{h}_{(iy,-x)}t} dt \, \eta_0$$

$$= \int_{0}^{\infty} e^{-t^4} \left(e^{\hat{h}_{(iy,-x)}t} + e^{\hat{h}_{(-iy,-x)}t} \right) dt \, \eta_0$$

$$= 2 \sum_{n=0}^{\infty} (-1)^n g_{2n}(1) a_{2n}(x, y), \qquad (7.10.17)$$

with

$$a_0(x, y) = 1, \qquad a_1(x, y) = y,$$
$$a_n(x, y) = \frac{1}{n} \left(y \, a_{n-1}(x, y) + 2 \, x \, a_{n-2}(x, y) \right), \qquad (7.10.18)$$

which is reconciled with our previous result, in terms of two-
variable Hermite polynomials, provided that one recognizes

$$J(1, x, -iy) = \sum_{n=0}^{\infty} \frac{(-1)^n}{n!} g_n(1) \left[H_n(-iy, -x) + H_n(iy, -x) \right].$$

$$(7.10.19)$$

106 A further important application is the use of the method within the framework of *evolutive PDE*, so let, $\forall x, \alpha, \beta \in \mathbb{R}, \forall t \in \mathbb{R}_0^+$,

$$\begin{cases} \partial_t F(x,t) = \left(\alpha \partial_x + \beta \partial_x^2\right) F(x,t) \\ F(x,0) = f(x), \end{cases} \tag{7.10.20}$$

whose (formal) solution is easily obtained as

$$F(x,t) = e^{(\alpha t)\partial_x + (\beta t)\partial_x^2} f(x). \tag{7.10.21}$$

The use of the formalism developed so far allows to write the r.h.s. of Eq. (7.10.21) in the form

$$F(x,t) = e^{\hat{h}_{(\alpha t,\ \beta t)}\partial_x} f(x)\eta_0, \tag{7.10.22}$$

by the use of standard exponential rules we obtain

$$F(x,t) = f\left(x + \hat{h}_{(\alpha t,\ \beta t)}\right)\eta_0, \tag{7.10.23}$$

which is still a formal solution unless we provide a meaning for the r.h.s. of Eq. (7.10.23). Let us therefore use the Fourier transform method to write

$$\begin{aligned} f\left(x + \hat{h}_{(\alpha t,\ \beta t)}\right)\eta_0 &= \frac{1}{\sqrt{2\pi}} \int_{-\infty}^{\infty} \tilde{f}(k) e^{ikx + ik\hat{h}_{(\alpha t,\ \beta t)}} dk\ \eta_0 \\ &= \frac{1}{\sqrt{2\pi}} \int_{-\infty}^{\infty} \tilde{f}(k) e^{ik(x+\alpha t) - k^2\beta t} dk, \end{aligned} \tag{7.10.24}$$

which is a kind of Gabor transform [93].

107 Let us now specialize the previous result to the case $f(x) = x^n$ and write

$$\begin{aligned} f\left(x + \hat{h}_{(\alpha t,\ \beta t)}\right)\eta_0 &= \left(x + \hat{h}_{(\alpha t,\ \beta t)}\right)^n \eta_0 \\ &= \sum_{s=0}^{n} \binom{n}{s} x^{n-s} \hat{h}_{(\alpha t,\ \beta t)}^s \eta_0 \\ &= H_n(x + \alpha t,\ \beta t), \end{aligned} \tag{7.10.25}$$

which is just the derivation from a different point of view of the following operational identity [18]:

$$e^{\kappa\partial_x+\lambda\partial_x^2}x^n = H_n(x+\kappa,\lambda), \qquad \forall x,\kappa,\lambda \in \mathbb{R}. \quad (7.10.26)$$

108 The possibilities for the applicability of the integration method discussed in this section arise if, inside the integrand, an exponential generating function is recognized. To clarify this point, we note that the integral

$$f(a,b,c) = \int_{-\infty}^{\infty} e^{-ax^2+\sqrt{x^2+bx+c}}dx, \qquad b^2-4c < 0, \quad a > 1, \quad (7.10.27)$$

can be written as

$$f(a,b,c) = \sqrt{\frac{\pi}{a}}e^{\frac{\hat{R}^2}{4a}} = \sqrt{\frac{\pi}{a}}\sum_{n=0}^{\infty}\frac{1}{n!}\left(\frac{1}{4a}\right)^n R_{2n}(b,c), \quad (7.10.28)$$

provided that

$$e^{\sqrt{x^2+bx+c}} = e^{\hat{R}x} = \sum_{n=0}^{\infty}\frac{x^n}{n!}\hat{R}^n, \qquad \hat{R}^n = R_n(b,c), \quad (7.10.29)$$

where $R_n(b,c)$ are polynomials of the parameter b,c. The following integral definition:

$$R_m(b,c) = \frac{m!}{2\pi}\int_0^{2\pi} e^{-im\phi}e^{\sqrt{e^{2i\phi}+be^{i\phi}+c}}d\phi, \quad (7.10.30)$$

has been used to benchmark the identity (7.10.28), with the full numerical integration of (7.10.27).

109 Laguerre–Schrödinger Equation

Introduce the Laguerre Coherent states[z]

$$|\Psi(t)\rangle = \sum_{n=0}^{\infty}\frac{(\hat{c}\,\Omega\,t)^n}{\sqrt{n!}}e^{-\frac{(\hat{c}\,\Omega\,t)^2}{2}}\varphi_0|\,n\,\rangle. \quad (7.10.31)$$

---

[z]Recall Section 7.5 and treat the same problem translated to a Laguerre–Schrödinger equation, which yields the result in Eq. (7.10.31) (see Eqs. (7.10.10)–(7.10.12)).

and show that the probability of finding the state $|\Psi(t)\rangle$ into an "n" state is given by the square amplitude reported as follows:

$$| \langle n|\Psi(t) \rangle |^2 = \frac{(\hat{c}\,\Omega\,t)^{2n}}{n!}e^{-(\hat{c}\,\Omega\,t)^2}\varphi_0 = \frac{X^n}{n!}\iota e_{2n}^{(2)}\,(-X),$$

(7.10.32)

where $X = (\Omega\,t)^2$. It is a normalized *Laguerre-Poisson* "*distribution*". Show that its mean value and variance, are respectively, given by

$$\langle n \rangle = \frac{X}{2}, \qquad\qquad \sigma_n^2 = \frac{1}{4!}\,(12 - 5X)\,X.$$

(7.10.33)

The quotes are due to the fact that Eq. (7.10.32) cannot be considered a probability distribution because it is not positively defined for all x values. This last statement may appear rather surprising, since we have defined a square amplitude. It should be however understood that the "square" of the scalar product in Eq. (7.10.32) is not a number but an operator, specified by its action on φ_0 and therefore it is not necessarily positively defined.

7.10.3   *Free Electron Laser*

Free Electron Laser (*FEL*) devices can be framed within the family of coherent radiation devices, originated by the Klystron. The *FEL* is a device capable of transforming the kinetic energy of a beam of electrons into electromagnetic radiation with laser-like properties. The *FEL* dynamics in the unsaturated regime is ruled by an equation of the type [41, 44, 114]

$$\partial_\tau\,a = i\,\pi\,g_0 \int_0^\tau \tau' e^{-i\,\nu\,\tau' - \frac{(\pi\mu_\varepsilon\tau')^2}{2}}\,a(\tau - \tau')\,d\tau', \qquad a(0) = 1,$$

(7.10.34)

where a represents the laser field amplitude, g_0 the small signal gain coefficient, ν is linked to the laser frequency and the coefficient μ_ε is a parameter regulating the effects of the inhomogeneous broadening due to the electrons' energy distribution.

It is an integro-differential equation of Volterra type. The kernel of the integral part is not trivial, Eq. (7.10.34) cannot indeed be solved analytically, unless $\mu_\varepsilon = 0$, as shown later. We compare different forms of solutions, based on perturbative methods, including the concepts relevant to the Hermite calculus developed in the previous sections.

In order to provide an idea of the point of view we are going to develop, we can write Eq. (7.10.34) in a slightly different form, more suitable for our purposes,

$$\partial_\tau a = \hat{V}(\tau)\, a, \qquad \hat{V}(\tau) = i\,\pi\, g_0 \int_0^\tau \tau' e^{-i\,\nu\,\tau' - \tau' \partial_\tau - (\pi\mu_\varepsilon)^2 \tau'^2}\, d\tau'.$$

(7.10.35)

The solution can accordingly be obtained by the iteration

$$\partial_\tau a_n(\tau) = \hat{V}^n a_0, \qquad a(\tau) = \sum_{n=0}^\infty a_n(\tau), \qquad a_0 = a(0) = 1,$$

(7.10.36)

where the *nth* term of the previous expansion (namely the Volterra series) can also be cast in the explicit form [85, 92]

$$\partial_\tau a_n = i\,\pi\, g_0 \int_0^\tau \tau' e^{-i\,\nu\,\tau' - \frac{(\pi\mu_\varepsilon \tau')^2}{2}}\, a_{n-1}(\tau - \tau')\, d\tau', \qquad a_0 = 1.$$

(7.10.37)

For later convenience, we call n the principal expansion index.

Even though the method is efficient and converges fast, the integrals cannot be done analytically but require a numerical treatment, which increases the computation times when a larger number of term is involved. We show how the wise use of generalized forms of multivariable polynomials opens new avenues for the solution of Eq. (7.10.34) and of Volterra-type equations more in general. A way to overcome the computation of integrals in analytical terms is provided by two-variable *HP* (1.2.5) which, by the use of the relevant generating function (1.2.6), yields

$$\partial_\tau a_n = i\pi g_0 \sum_{m=0}^\infty \frac{1}{m!} \int_0^\tau \tau'^{(m+1)}\, H_m\left(-i\nu, -\frac{(\pi\mu_\varepsilon)^2}{2}\right) a_{n-1}(\tau - \tau') d\tau'.$$

(7.10.38)

The Hermite expansion index m is therefore nested into the principal index n. The integration procedure is now straightforward and we get

(1) **First-order solution**

$$a_1 = i\,\pi\,g_0 \sum_{m=0}^{\infty} \frac{H_m\left(-i\,\nu, -\frac{(\pi\,\mu_\varepsilon)^2}{2}\right)}{m!} \int_0^\tau d\tau' \int_0^{\tau'} \tau''^{(m+1)} d\tau'',$$

(7.10.39)

which yields[aa]

$$a_1 = i\,\pi\,g_0 \sum_{m_1=0}^{\infty} \alpha_{m_1} \tau^{m_1+3}, \qquad \alpha_{m_1} = \frac{H_{m_1}}{m_1!(m_1+2)(m_1+3)}.$$

(7.10.40)

(2) **Second-order solution**

$$a_2 = (i\,\pi\,g_0)^2 \sum_{m_2=0}^{\infty} \frac{H_{m_2}}{m_2!} \int_0^\tau d\tau' \int_0^{\tau'} \tau''^{(m_2+1)} a_1(\tau' - \tau'')d\tau'',$$

(7.10.41)

which after some algebra yields

$$a_2 = (i\,\pi\,g_0)^2 \sum_{m_1,m_2=0}^{\infty} \alpha_{m_1,m_2} \tau^{m_1+m_2+6},$$

$$\alpha_{m_1,m_2} = \alpha_{m_1} \sum_{s=0}^{m_1+3} \binom{m_1+3}{s} \frac{H_{m_2}}{m_2!(m_2+s+2)(m_2+m_1+6)}$$

$$= \frac{(-1)^s\,H_{m_1}H_{m_2}}{m_2!m_1!(m_1+2)(m_1+3)(m_2+m_1+6)}$$

$$\cdot \sum_{s=0}^{m_1+3} \binom{m_1+3}{s} \frac{(-1)^s}{(m_2+s+2)}.$$

(7.10.42)

We can use the previous two contributions to check whether the results we obtain concerning, e.g., the FEL gain agreement with analogous conclusions already given in the literature.

[aa]The arguments of the Hermite have been omitted to avoid cumbersome expressions.

(3) **Higher-order solution**

The interesting aspect of the present nested procedure is that the nth order can be computed in a modular way just looking at the symmetries of the expansion itself. The nth order term indeed reads

$$a_n = (i \pi g_0)^n \sum_{m_1,..m_n=0}^{\infty} \alpha_{m_1,..m_n} \tau^{\left(\sum_{r=1}^n m_r + 3n\right)},$$

$$\alpha_{m_1,..m_n} = \alpha_{m_1,..m_{n-1}} \sum_{s=0}^{\left(\sum_{r=1}^{n-1} m_r + 3(n-1)\right)} \binom{\sum_{r=1}^{n-1} m_r + 3(n-1)}{s}$$

$$\cdot \frac{(-1)^s H_{m_n}}{m_n!(m_n + s + 2)\left(\sum_{r=1}^n m_r + 3n\right)}. \tag{7.10.43}$$

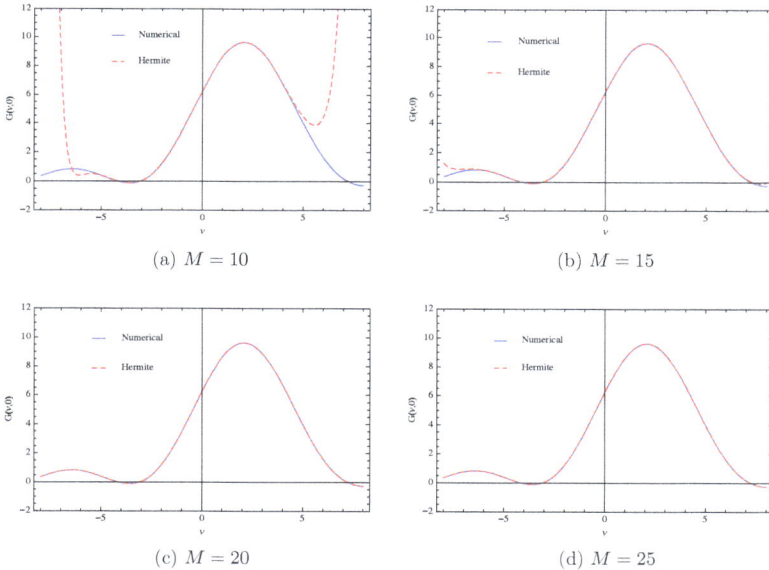

(a) $M = 10$ (b) $M = 15$

(c) $M = 20$ (d) $M = 25$

Figure 7.5: Comparison between complete numerical integration with no broadening effects $(\mu_\varepsilon = 0)$ $G(\nu, 0) = \| a \|^2 - 1$ with $g_0 = 5$ at the end time $\tau = 1$, performed by Mathematica, and Hermite solution $G(\nu, 0) = \|a_0 + a_1 + a_2\|^2 - 1$ at different truncation levels $(M = 10, M = 15, M = 20, M = 25)$.

The square modulus of the function a at $\tau = 1$ reads namely[bb]

$$G(\nu, \mu_\varepsilon) = \|a\|^2 - 1. \tag{7.10.44}$$

The derivation of

$$G(\nu, \mu_\varepsilon) = \|a_0 + a_1 + a_2 + \cdots + a_n\|^2 - 1 \tag{7.10.45}$$

requires the calculation of the norm of the quadratic sum of $n + 1$ terms, of which one is constant (unity for a_0) and n, according to the method outlined in the previous section, are the HP expansion as in Eqs. (7.10.40), (7.10.42) and (7.10.43). In Fig. 7.5, a second-order gain function is reported, where the present procedure (at different

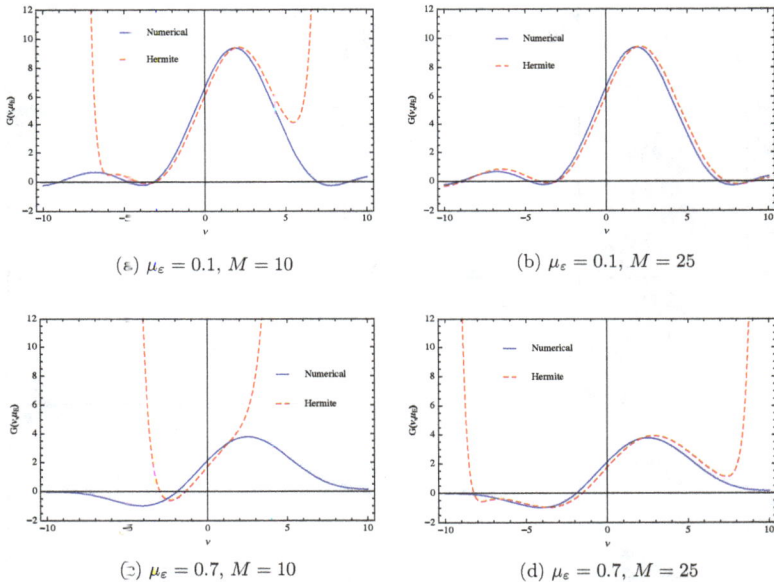

(a) $\mu_\varepsilon = 0.1$, $M = 10$

(b) $\mu_\varepsilon = 0.1$, $M = 25$

(c) $\mu_\varepsilon = 0.7$, $M = 10$

(d) $\mu_\varepsilon = 0.7$, $M = 25$

Figure 7.6: Comparison between complete numerical integration with different broadening effects ($\mu_\varepsilon = 0.1, \mu_\varepsilon = 0.7$) $G(\nu, \mu_\varepsilon) = \|a\|^2 - 1$ with $g_0 = 5$ at the end time $\tau = 1$, performed by Mathematica, and Hermite solution $G(\nu, \mu_\varepsilon) = \|a_0 + a_1 + a_2\|^2 - 1$ at different truncation levels ($M = 10, M = 25$).

[bb]We have subtracted the initial condition for convenience.

truncation levels) is confronted with those from a complete numerical integration and no differences are foreseeable. The truncation of the Hermite expansion obviously affects the speed and the approximation of the gain curve. The method allows also taking into account the broadening effect as shown in Fig. 7.6. It has to be noted that for higher broadening effects a higher truncation level is needed to achieve a good approximation, as shown in Figs. 7.6c–7.6d.

7.10.4 *Bethe–Salpeter Equation and Matrix Environment*

110 The use of pseudo-differential operators for the solution of *Klein–Gordon or Bethe–Salpeter equations* has gained considerable interest in the last few years, we use the methods discussed in the previous sections to study the solution of the stationary Klein–Gordon equation describing the axion photon coupling in a transverse magnetic field. According to the analysis developed in [144], photon γ and axion a fields can be viewed as two-state polarization, which, propagating in an intense magnetic field, undergoes a kind of Cotton–Mouton rotation. Within such a framework the photon–axion interaction can be viewed as a kind of Primakoff process (see Fig. 7.7) in which the vertex coupling occurs between a real external photon, the virtual photon associated with the static magnetic field and the axion.

The equation we consider is [144]

$$\left(\frac{\partial}{\partial t}\right)^2 \begin{pmatrix} \gamma \\ a \end{pmatrix} = -\omega^2 \begin{pmatrix} 2n-1 & \dfrac{g_{a\gamma}B}{\omega} \\ \dfrac{g_{a\gamma}B}{\omega} & 1 - \dfrac{m_a^2}{\omega^2} \end{pmatrix} \begin{pmatrix} \gamma \\ a \end{pmatrix}, \quad (7.10.46)$$

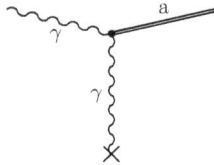

Figure 7.7: Primakoff process for photon (γ)–photon (γ)–axion (a) production.

where n is the refractive photon index associated with the magnetic field B, $g_{a\gamma}$ the axion photon coupling constant, m_a is the axion mass and ω is the external photon frequency. Without further considering the specific details, we find that the eigenvalues of the matrix are

$$\lambda_\pm = \frac{N + M \pm \sqrt{[N - M]^2 + 4G^2}}{2},$$

$$N = 2n - 1, \qquad M = 1 - \left(\frac{m_a}{\omega}\right)^2, \qquad G = \frac{g_{a\gamma}B}{\omega}. \tag{7.10.47}$$

If we make the assumption that $\frac{m_a}{\omega} \ll 1$ (which means that the energy of the external photon is much larger than the axion mass [135]) and $N \cong 1$, we are left with $\lambda_\pm = 1 \pm G$. Furthermore, by noting also that $G \ll 1$, we find

$$f_0(\lambda_+, \lambda_-) = \frac{1 - G^2}{2G}\left(\frac{1}{\sqrt{1 - G}} - \frac{1}{\sqrt{1 + G}}\right) \cong \frac{1}{2} - \frac{3}{16}G^2,$$

$$f_1(\lambda_+, \lambda_-) = \frac{\sqrt{1 - G} - \sqrt{1 + G}}{-2G} \cong \frac{1}{2} + \frac{G^2}{16}, \tag{7.10.48}$$

and

$$e^{i\omega\, t\, \sqrt{\hat{A}}} = e^{-i\omega t[(1 - \frac{1}{8}G^2)]}\begin{pmatrix} \cos(f_1 G \omega\, t) - \sin(f_1 G \omega\, t) \\ \sin(f_1 G \omega\, t) \quad \cos(f_1 G \omega\, t) \end{pmatrix}, \tag{7.10.49}$$

which is essentially a rotation matrix induced by the axion-photon coupling constant. The validity of our solution is limited to the case in which both $\lambda_\pm$ are non-negative, therefore the following conditions are to be satisfied:

$$m_a \leq \omega\sqrt{1 - \frac{1}{2n - 1}\left(\frac{g_{a\gamma}B}{\omega}\right)^2}. \tag{7.10.50}$$

It is interesting to look at the behavior vs time of the probability of creating an axion during the interaction for

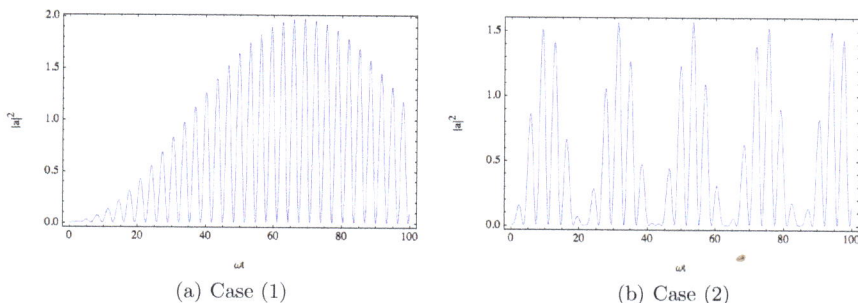

(a) Case (1) (b) Case (2)

Figure 7.8: Axion generation probability ($| \ a \ |^2 \cdot 10^{15}$) vs ωt for different parameters.

different values of the parameter. In Fig. 7.8, we have shown $|a|^2$ vs $\omega \, t$ for the cases

(a) $\dfrac{g_{a\gamma}B}{\omega} \cong 2 \cdot 10^{-9}$, $\quad \left(\dfrac{m_a}{\omega}\right) \cong 0.3$;

(b) $\dfrac{g_{a\gamma}B}{\omega} \cong 10^{-8}$, $\quad \left(\dfrac{m_a}{\omega}\right) \cong 0.71$.

111 Regarding the extension of the method to matrices with dimensionality higher than 3×3, we note that in the case of a 4×4 matrix we have

$$e^{\tau \sqrt{\hat{A}}} = \prod_{r=0}^{3} e^{f_r \tau \, \hat{A}^r}. \tag{7.10.51}$$

The use of the Airy transform [18] allows to write

$$e^{\alpha \hat{A}^3} = \int_{-\infty}^{\infty} Ai(t) \, e^{\sqrt[3]{\alpha}t \, \hat{A}} dt, \qquad Ai(t) = \frac{1}{2\pi} \int_{-\infty}^{\infty} e^{\left(\frac{i}{3}\xi^3 + it\xi\right)} d\xi. \tag{7.10.52}$$

Thus, finally getting, for the square root of a 4×4 matrix, the following expression in terms of Airy and Gauss transforms:

$$e^{\tau \sqrt{\hat{A}}} = \frac{e^{f_0(\tau)}}{\sqrt{\pi}}$$

$$\cdot \int_{-\infty}^{+\infty} \int_{0}^{\infty} e^{-\xi^2} Ai(\lambda) \hat{U}(f_1 \tau + 2\sqrt{f_2 \tau}\xi + \sqrt[3]{f_3 \tau}\lambda) d\xi d\lambda,$$

$$\hat{U}(\tau) = e^{\tau \hat{A}}. \tag{7.10.53}$$

The case of matrices with higher-order dimensionalities $(5 \times 5, \ldots)$ requires only higher-order integral transforms of the type discussed in [20].

The use of the Lévy–Smirnov distribution allows to write (see Eq. (5.5.23))

$$\exp\left\{-\tau \sqrt{\begin{pmatrix} 0 & -\omega_3 & \omega_2 \\ \omega_3 & 0 & -\omega_1 \\ -\omega_2 & \omega_1 & 0 \end{pmatrix}}\right\} = \int_0^\infty g_{\frac{1}{2}}(\eta)\, \hat{R}(-\eta\,\tau^2) d\eta.$$

(7.10.54)

Taking into account that

$$\int_0^\infty g_{\frac{1}{2}}(\eta)\, d\eta = 1,$$

$$\int_0^\infty g_{\frac{1}{2}}(\eta) e^{-i\eta x}\, d\eta = e^{-\frac{\sqrt{2x}}{2}}\left(\cos\left(\frac{\sqrt{2x}}{2}\right) - i\sin\left(\frac{\sqrt{2x}}{2}\right)\right)$$

(7.10.55)

and, by using the explicit form of the Rodrigues matrix (see Eqs. (5.5.22) and (5.5.23)) we end up with

$$e^{-\tau\sqrt{\hat{A}}} = \hat{1} - \frac{e^{-\frac{\sqrt{2\Omega}}{2}\tau}}{\Omega}\sin\left(\frac{\sqrt{2\Omega}}{2}\tau\right)\hat{A}$$

$$+ \frac{1}{\Omega^2}\left(1 - e^{-\frac{\sqrt{2\Omega}}{2}\tau}\cos\left(\frac{\sqrt{2\Omega}}{2}\tau\right)\right)\hat{A}^2.$$

(7.10.56)

The possibility of extending the method to higher-order roots follows the same procedure we have envisaged and, for example,

$$e^{-\tau\sqrt[n]{\hat{A}}} = \int_0^\infty g_{\frac{1}{n}}(\eta)\, e^{-\eta\tau^n \hat{A}} d\eta,$$

(7.10.57)

where the Lévy stable function $g_{\frac{1}{n}}(\eta)$ is in general expressible in terms of *Meijer G-functions* (see [142]).

This last exercise closes the Chapter (and the book!). We have provided a large number of examples which cover only a part of the possibilities offered by the method... Further examples can be found in the quoted literature.

Conclusions

This book considered various aspects of *Umbral Calculus* applied to many different topics in Pure and Applied Mathematics. We have used the term umbral even though it is not exactly the same formalism proposed by Rota and coworkers. We have stressed that the point of view developed here shares even more analogies with the Heaviside symbolic calculus and preserves the relevant spirit, because it has been conceived for applications.

The most significant benefits, introduced by such a technique, are associated with the simplifications it provides in analytical computations, and in the possibility it yields a thread between seemingly uncorrelated topics. We have indeed treated such desparate points, ranging from Fractional derivatives, Bessel functions, Special polynomials and Harmonic numbers using always the same formalism which has been exploited as a tool kit to disclose the relevant properties of different mathematical entities.

Although the book is rather long, we have been obliged to leave out many interesting subjects and applications. In particular, we did not mention the theory of multivariable Bessel functions which, within the present context, acquires a particularly tasty flavor allowing noticeable simplifications for the study of problems associated, e.g., with the treatment of synchrotron radiation and *FEL*.

We hope that the form we have presented yields an idea of the "wildly wide" applicability of the method and of the possibility it may offer.

Bibliography

[1] M. Abramovitz and I.A. Stegun (Eds.), *Handbook of Mathematical Functions with Formulas, Graphs and Mathematical Tables*, Chapter 9, 9th printing, New York: Dover, 1972, p. 362, Eq. (9.1.57).

[2] M. Abramovitz and I.A. Stegun (Eds.), Parabolic cylinder function, *Handbook of Mathematical Functions with Formulas, Graphs and Mathematical Tables*, Chapter 19, 9th printing, New York: Dover, 1972, pp. 685–700.

[3] L.C. Andrews, *Special Functions For Engeneers and Applied Mathematicians*, New York: Mc Millan, 1985.

[4] A.H. Ansari, X. Liu, and V.N. Mishra, On Mittag-Leffler function and beyond, *Nonlinear Science Letters A*. 8(2) (2017), 187–199.

[5] P. Appéll, Sur une classe de polynomes, *Annales Scientifiques de l'École Normale Supérieure, 2e Série*. 9 (1880), 119–144.

[6] P. Appéll and J. Kampé de Fériét, *Fonctions Hypergeometriques and Hyperspheriques. Polynomes d'Hermite*, Paris: Gauthiers-Villars, 1926.

[7] M. Artioli and G. Dattoli, Geometry of two-variable Legendre polynomials, From Wolfram Demonstrations Project-A From Wolfram Web Resource, demonstrations.wolfram.com/GeometryOfTwo VariableLegendrePolynomials.

[8] M. Artioli and G. Dattoli, *The Geometry of Hermite Polynomials*, From Wolfram Demonstrations Project, Mar. 2015, http://demonst rations.wolfram.com/TheGeometryOfHermitePolynomials/.

[9] M. Artioli, G. Dattoli, and S. Licciardi, *Motzkin numbers and their geometrical interpretation*, From Wolfram Demonstrations Project, 2017.

[10] M. Artioli, G. Dattoli, S. Licciardi, and S. Pagnutti, Motzkin numbers: an operational point of view, *J. Integer Seq.*, 21 (2018), Article 18.7.5, cited on Online Electronic Integer Sequences as arXiv:1703.07262.

[11] m. Artioli, G. Dattoli, S. Licciardi, and S. Pagnutti, Fractional Derivatives, Memory kernels and solution of Free Electron Laser Volterra type equation, *Mathematics*, 5(4) (2017), 73; doi: 10.3390/math5040073. Selected for Special Issue Cover.

[12] M. Artioli *et al.*, *A 250 GHz Radio Frequency CARM Source for Plasma Fusion*, Conceptual Design Report, ENEA, 2016, p. 154, ISBN: 978-88-8286-339-5.

[13] D. Babusci and G. Dattoli, *On Ramanujan Master Theorem*, arXiv:1103.3947 [math-ph].

[14] D. Babusci, G. Dattoli, and M. Carpanese, *The Algebra of Factorial Polynomials*, (2011), arXiv:1107.4024v1 [math.AP].

[15] D. Babusci, G. Dattoli, E. Di Palma, and E.N. Petropoulou, The Humbert-Bessel functions, Stirling numbers and probability distributions in coincidence problems, *Far East J. Math. Sci.*, in printing.

[16] D. Babusci, G. Dattoli, K. Gorska, and K. Penson, Lacunary Generating Functions for Laguerre Polynomials, *Seminaire Lotharingien de Combinatoire* (2017), Article B76b.

[17] D. Babusci, G. Dattoli, K. Gorska, and K.A. Penson, The spherical Bessel and Struve functions and operational methods, *Appl. Math. Comput.*, 238(1) (2014), 1–6.

[18] D. Babusci, G. Dattoli, S. Licciardi, and E. Sabia, *Mathematical Methods for Physicists*, Singapore: World Scientific, 2019.

[19] D. Babusci, G. Dattoli, and M. Quattromini, Relativistic equations with fractional and pseudodifferential operators, *Phys. Rev. A 83* (2011), 062109; Kowalski and J. Rembielinski, Salpeter equation and probability current in the relativistic quantum mechanics, *Phys. Rev. A 84* (2011), 012108.

[20] D. Babusci, G. Dattoli, and D. Sacchetti, The Airy transform and the associated polynomials, *centr.eur.j.phys.* 9 (2011), 1381, doi:10.2478/s11534-011-0057-9.

[21] E. Barkai, Fractional Fokker-Planck equation, solution, and application, *Phys. Rev. E 63* (2001), 046118.

[22] C. Baumgarten, Use of real Dirac matrices in two-dimensional coupled linear optics, *Physical Review Special Topics - Accelerators and Beams*, 14 (2011), 114002.

[23] G. Baym, *Lectures on Quantum Mechanics*, W.A. Benjamin, 1969.

[24] E.T. Bell, The History of Blissard's Symbolic Method, with a Sketch of its Inventor's Life, *The American Mathematical Monthly*, 45(7) (1938), 414–421.

[25] B.C. Berndt, *Ramanujan's Notebooks: Part I*, New York: Springer-Verlag, 1985, p. 298.

[26] M.V. Berry and C.J. Howls, Integrals with coalescing saddles, in: *NIST Handbook of Mathematical Functions*, Chapter 36, Cambridge: Cambridge University Press, 2010, pp. 775–793.

[27] G. Birkhoff and S. Mac Lane, *A Survey of Modern Algebra*, 3rd edn., Chapter X, Macmillan, 1965, § 6.

[28] P. Blasiak, G. Dattoli, A. Horzela, K.A. Penson and K. Zhukovsky, Motzkin numbers, central trinomial coefficients and hybrid polynomials, *J. Integer Seq.* 11 (2008), Art. 08.1.1.

[29] J. Blissard, Theory of Generic Equations, *Quarterly Journal of Pure and Applied Mathematics*, 4 (1861), 279–305; 5 (1862), 58–75, 185–208.

[30] R.P. Boas and R.C. Buck, Polynomial expansions of analytic functions, in: *Ergebnisse der Mathematik und ihrer Grenzgebiete. Neue Folge*, 19, Berlin, New York: Springer-Verlag, MR 0094466, 1958.

[31] J. Bohacik, P. Augustin, and P. Presnajder, Non-perturbative anharmonic correction to Mehler's presentation of the harmonic oscillator propagator, *Ukr. J. Phys.* 59 (2014), 179.

[32] G. Boole, *Calculus of Finite Differences*, 2nd edn., Cambridge: MacMillan, 1872.

[33] M. Born and E. Wolf, *Principles of Optics*, Pergamon Press, 1970.

[34] Y.A. Brychkov, On multiple sums of special functions, *Int. Trans. Spec. F.* 21(12) (2010), 877–884.

[35] P. Cartier, "Mathemagics" (A tribute to L. Euler and R. Feynman), *Seminaire Lotharingien de Combinatoire* 44 (2000), Article B44d.

[36] Y.B. Cheick, Decomposition of Laguerre polynomials with respect to the cyclic group, *J. Comp. Appl. Math.*, 99 (1998), 55–66.

[37] T.S. Chihara, *An Introduction to Orthogonal Polynomials*, Mineola, New York: Dover Pub. Inc., 2011.

[38] F.M. Cholewinski, The Finite Calculus associated with Bessel Functions, *J. Amer. Math. Soc.*, 75 (1988).

[39] F.M. Cholewinski and J.A. Reneke, The Generalized Airy Diffusion Equation, *Electronic J. Diff. Equs.*, 87 (2003), 1–64, ISSN: 1072-6691.

[40] F. Ciocci, G. Dattoli, L. Giannessi, A. Torre, and G. Voykov, Analytic and Numerical Study of two frequency undulator radiation, *Phys. Rev. E*, 47(3) (1993).

[41] F. Ciocci, A. Torre, and G. Dattoli, *Insertion Devices for Synchrotron Radiation and Free Electron Laser*, River Edge, NJ: World Scientific, 2000.

[42] D. Cocolicchio, G. Dattoli, and H.M. Srivastava (Eds.), *Advanced special functions and applications*, (Melfi, 1999), *Proc. Melfi Sch. Adv. Top. Math. Phys.* 1(2000), 147–164.

[43] M.W. Coffee, Expressions for Harmonic Number Generating Functions, in: T. Amdeberhan, E.T. Simos, V.H. Moll (Eds.), *Contemporary Mathematics, 517, Gems In Experimental Mathematics*, AMS Special Session Experimental Mathematics, 2009.

[44] W.B. Colson, *Classical Free Electron Laser theory, Laser Handbook*, vol. VI by W. B. Colson, C. Pellegrini and A. Renieri, Amsterdam: North Holland, 1990.

[45] J.H. Conway and R.K. Guy, *The Book of Numbers*, New York: Copernicus, 1996.

[46] D. Cvijović, The Dattoli-Srivastava Conjectures Concerning Generating Functions Involving the Harmonic Numbers, *Appl. Math. Comput.*, 215(9) (2010), 4040–4043.

[47] G. Dattoli, Generalized polynomials, operational identities and their applications, *J. Comput. Appl. Math.*, 118(12) (2000), 111–123.

[48] G. Dattoli, Hermite-Bessel and Laguerre-Bessel functions: a by-product of the monomiality principle, in: D. Cocolicchio, G. Dattoli and H.M. Srivastava (Eds.), *Advanc. Special Functions and Applications (Melfi, 1999)* Rome: Aracne Editrice, 2000, pp. 147–164.

[49] G. Dattoli, Laguerre and generalized Hermite polynomials: the point of view of the operational method, *Int. Trans. Spec. F.*, 15(2) (2004).

[50] G. Dattoli and M. Artioli, *Vedic Math, Magic Math,...Damn Math*, RT/2015/6/ENEA.

[51] G. Dattoli and M. Del Franco, The Euler legacy to modern physics, *Lecture Notes of Seminario Interdisciplinare di Matematica* 9 (2010), 1–24.

[52] G. Dattoli, E. Di Palma, S. Licciardi, and E. Sabia, From circular to Bessel functions: a transition through the umbral method, *Fractal Fract.*, 1(1) (2017), 9, doi:10.3390/fractalfract1010009.

[53] G. Dattoli, E. Di Palma, F. Nguyen, and E. Sabia, Generalized trigonometric functions and elementary application, *Int. J. Appl. Comput. Math* (2016), doi:10.1007/s40819-016-0168-5.

[54] G. Dattoli, E. Di Palma, E. Sabia, K. Gorska, A. Horzela, and K.A. Penson, Operational versus umbral methods and the Borel transform, *Int. J. Appl. Comput. Math.* (2017), 1–22.

[55] G. Dattoli, E. Di Palma, E. Sabia, and S. Licciardi, Products of Bessel functions and associated polynomials, *Applied Mathematics and Computation*, 266(C), (2015), 507–514.

[56] G. Dattoli, E. Di Palma, E. Sabia, and S. Licciardi, Quasi exact solution of the Fisher equation, *Appl. Math.*, 4(8A) (2013), 7–12, doi: 10.4236/am.2013.48A002.

[57] G. Dattoli, J.C. Gallardo, and A. Torre, An algebraic view to the operatorial ordering and its applications to optics, *Riv. Nuovo Cimento*, 11(3) (1988) 1–79.

[58] G. Dattoli, B. Germano, S. Licciardi, and M.R. Martinelli, Integrals of special functions and umbral methods, *Int. J. Appl. Comput. Math. (IACM)*, 7, 120, 2021.

[59] G. Dattoli, B. Germano, S. Licciardi, and M.R. Martinelli, Hermite calculus, in: J. Gielis, P. Ricci, I. Tavkhelidze (Eds.), *Modeling in Mathematics, Atlantis Transactions in Geometry*, vol. 2, Paris, Springer: Atlantis Press, 2017, pp. 43–52.

[60] G. Dattoli, B. Germano, S. Licciardi, and M.R. Martinelli, On an umbral treatment of Gegenbauer, Legendre and Jacobi polynomials, *Int. Math. Forum*, 12(11) (2017), 531–551.

[61] G. Dattoli, B. Germano, S. Licciardi, and M.R. Martinelli, Umbral methods and harmonic numbers, *Axioms*, 7(3) (2018), 62. Special Issue *Math. Analy. Appl.*, doi.org/10.3390/axioms7030062.

[62] G. Dattoli, B. Germano, M.R. Martinelli, and P.E. Ricci, A novel theory of Legendre polynomials, *Math. Comput. Modell.*, 54 (2011), 80–87.

[63] G. Dattoli, B. Germano, M.R. Martinelli, and P.E. Ricci, Integral Transforms and Special Functions, 19(4) (2008), 259–266.

[64] G. Dattoli, B. Germano, M.R. Martinelli, and P.E. Ricci, Lacunary generating functions of Hermite polynomials and symbolic methods, *Ilirias J. Math.*, 4(1) (2015), 16–23, ISSN: 2334-6574, URL: http://www.ilirias.com.

[65] G. Dattoli, B. Germano, M.R. Martinelli, and P.E. Ricci, The negative derivative operator, *Integ. Transf. Spec. Funct.*, 19(3/4) (2008), 259–266.

[66] G. Dattoli, K. Gorska, A. Horzela, S. Licciardi, and R.M. Pidatella, Comments on the properties of Mittag-Leffler function, *Eur. Phys. J. Special Topics*, 226, (2017), 3427–3443. EDP Sciences, Springer-Verlag, 2018, doi.org/10.1140/epjst/e2018-00073-1.

[67] G. Dattoli, S. Khan, and P.E. Ricci, On Crofton-Glaisher type relations and derivation of generating functions for Hermite polynomials including the multi-index case, *Integr. Transf. Spec. Funct.* 19(1), (2008), 1–9.

[68] G. Dattoli and S. Licciardi, Operational, umbral methods, Borel transform and negative derivative operator techniques, *Integ. Transf. and Spec. Funct.*, 31(3) (2020), pp. 192–220, doi.org/10.1080/10652469.2019.1684487.

[69] G. Dattoli, S. Licciardi, F. Nguyen, and E. Sabia, Evolution equations involving matrices raised to non-integer exponents, *Model. in Math. Atlant. Trans. Geom.*, 2 (2017), 31–41.

[70] G. Dattoli, S. Licciardi, and R.M. Pidatella, Theory of generalized trigonometric functions: From Laguerre to airy forms, *J. Mathematical Anal. Appl.*, 468(1) (2018), 103–115, doi.org/10.1016/j.jmaa.2018.07.044.

[71] G. Dattoli, S. Licciardi, and E. Sabia, Generalized Trigonometric Functions and Matrix Parameterization, *Int. J. Appl. Comput. Math.*, (2017) 1–14, doi.org/10.1007/s40819-017-0427-0.

[72] G. Dattoli, S. Lorenzutta, and C. Cesarano, *From Hermite to Humbert Polynomials*, Vol. XXXV, Rend. Istit. Mat. Univ. Trieste, 2003, pp. 37–48.

[73] G. Dattoli, S. Lorenzutta, G. Maino, A. Torre, G. Voykov, and C. Chiccoli, Theory of two-index Bessel functions and applications to physical problems, *J. Math. Phys.*, 35 (1994), 3636.

[74] G. Dattoli, S. Lorenzutta, A.M. Mancho, and A. Torre, Generalized polynomials and associated operational identities, *J. Comput. Appl. Math.*, 108 (1999), 209–218.

[75] G. Dattoli, S. Lorenzutta, P.E. Ricci, and C. Cesarano, On a family of hybrid polynomials, *J. Integr. Transf. Spec. Funct.* 15 (2004), 485–490.

[76] G. Dattoli, M. Migliorati, and P.E. Ricci, *The Eisentein group and the pseudo hyperbolic function* (2010), arXiv:1010.1676 [math-ph].

[77] G. Dattoli, P.L. Ottaviani, A. Torre, and L. Vazquez, Evolution operator equations-integration with algebraic and finite-difference methods-applications to physical problems in classical and quantum mechanics and quantum field theory, *Rivista del Nuovo Cimento* 20 (1997), 1.

[78] G. Dattoli, A. Renieri, and A. Torre; *Lectures on Free Electron Laser Theory and Related Topics*, Singapore: World Scientific, 1993.

[79] G. Dattoli, P.E. Ricci, and L. Marinelli, *Generalized truncated exponential polynomials and applications*, Rendiconti dell'Istituto di Matematica dell'Università di Trieste, *An Int. J. Math.*, 34 (2002) 9–18.

[80] G. Dattoli and E. Sabia, Generalized transforms and special functions, (2010), RT/2009/42/ENEA, arXiv:1010.1679 [math-ph].

[81] G. Dattoli, E. Sabia, and M. Del Franco, *The pseudo-hyperbolic functions and the matrix representation of Eisenstein complex numbers*, (2010), arXiv:1003.2698v1 [math-ph].

[82] G. Dattoli and H.M. Srivastava, A note on harmonic numbers, umbral calculus and generating functions, *Appl. Math. Lett.*, 21(6) (2008), 686–693.

[83] G. Dattoli and A. Torre, Operatorial methods and two variable Laguerre polynomials, *Acc. Scienze Torino, Atti Sc. Fis.*, 1 (1998), 132.

[84] G. Dattoli, A. Torre, and M. Carpanese, Operational rules and arbitrary order hermite generating functions, *J. Math. Anal. Appl.*, 227(1) (1998), 98–111.

[85] G. Dattoli, A. Torre, C. Centioli, and M. Richetta, Free electron laser operation in the intermediate gain region, *IEEE J-QE*, 25 (1989), 2327.

[86] G. Dattoli *et al.*, Undulator and free electron laser radiation for fundamental physics research, talk at the *7th International Conference Charged & Neutral Particles Channeling Phenomena Channeling 2016*, https://agenda.infn.it/conferenceOtherViews.py/view=standard&confId=10663.

[87] E. Di Palma, E. Sabia, G. Dattoli, S. Licciardi, and I. Spassovsky, Cyclotron auto resonance maser and free electron laser devices: a unified point of view, *J. Plasma Phys.*, 83(1) (2017).

[88] G. Doetsch, *Handbuch der Laplace Transformation*, Basel: Birkhnauser, 1950–1956.

[89] J.S. Dowker, *Poweroids revisited, An old symbolic approach*, arXiv:1307.3150 [math.CO].

[90] D.E. Edmunds, P. Gurka, and J. Lang, Properties og generalized trigonometric functions, *J. Approx. Theory*, 164 (2012), 47–56.

[91] L. Ehrenpreis, The Borel Transform, in: T. Aoki, H. Majima, Y. Takei, and N. Tose (Eds.), *Algebraic Analysis of Differential Equations*, Berlin: Springer, 2008.

[92] For various material about FEL formulae, see also www.fel.enea.it.

[93] H.G. Feichtinger and T. Strohmer (Eds.), *Gabor Analysis and Algorithms Theory and Applications*, Birkhäuser, 1998.

[94] W. Feller, *An Introduction to Probability Theory and Its Applications*, vol. 2, 2nd edn., Wiley India, 2008.

[95] E. Ferrari, *Bollettino UMI*, 18B, 933 (1981). For early suggestions see also F.D. Bugogne, *Math. Comp.* 18 (1964), 314.

[96] P. Fjelstad and S.G. Gal, Two-dimensional geometries, topologies, trigonometries and physics generated by complex-type numbers, *Adv. Appl. Clifford Algebras*, 11 (2001), 81.

[97] J.W.L. Glaisher, On the Bernoullian Function, *Quart. J. Pure Appl. Math.*, 29 (1898), 1–168.

[98] F. Gori, Flattened gaussian beams, *Opt. Commun.*, 107(5-6) (1994), 335–341.

[99] K. Gorska, D. Babusci, G. Dattoli, G.H.E. Duchamp, and K.A. Penson, The Ramanujan master theorem and its implications for special functions, *Appl. Math. Comp.* 218 (2012), 11466–11471.

[100] K. Gorska, K.A. Penson, D. Babusci, G. Dattoli, and G.H.E. Duchamp, Operator solutions for fractional Fokker-Planck equations, *Pys. Rev. E* 85 (2012), 031138.

[101] R.W. Gosper, *Harmonic summation and exponential gfs.*, mathfun@cs.arizona.edu posting (1996).

[102] G.H. Hardy, Ramanujan, *Twelve Lectures on Subjects Suggested by his Life and Work*, Cambridge: Cambridge University Press, 1940.

[103] R. Hatz, M. Korpinen, V. Hanninen, and L. Halonen, Generalized intermolecular interaction tensor applied to long-range interactions in hydrogen and coinage metal (Cu, Ag, and Au) clusters, *J. Phys. Chem. A*, 119 (2015), 11729–11736.

[104] O. Heaviside, Electromagnetic induction and its propagation, *The Electrician*, 1887.

[105] R. Hermann, *Fractional Calculus: An Introduction for Physicists*, 2nd ed., Singapore: World Scientific, 2014.

[106] J.D. Jackson, *Classical Electrodynamics*, 3rd ed., Wiley, 1999.

[107] H.M. Jeffery, On a method of expressing the combination and homogeneous products of consecutive numbers and their powers by means of the differences of nothing, *Quart. J. Math.* 4 (1861), 364–377, https://gdz.sub.uni-goettingen.de/id/PPN600494829_0004?tify ={%22pages%22:[3],%22panX%22:0.616,%22panY%22:0.59,%22view %22:%22info%22,%22zoom%22:0.632}.

[108] H.M. Jeffery, On the expansion of powers of the trigonometrical ratios in terms of series of ascending powers of the variables, *Quart. J. Math.* 6 (1862), 91–108, https://gdz.sub.uni-goettingen.de/ id/PPN600494829_0005?tify= {%22pages%22:[5],%22panX%22:0.594, %22panY%22:0.788,%22view%22:%22info%22,%22zoom%22:0.442}.

[109] R.A. Horn and C.R. Johnson, *Topics in Matrix Analysis*, Cambridge University Press, 1991 See Section 6.1.

[110] P. Humbert, Some extensions of Pincherle's polynomials, *Proc. Edinburgh Math. Soc.*, 39 (1921), 21–24.

[111] J. Jansen, *Rotations in three four and five dimensions*, arXiv:1103 .5263v1[math.MG].

[112] S. Khan and A.A. Al-Gonah, Operational methods and Laguerre - Gould-Hopper polynomials, *Appl. Math. Comput.*, 218(19) (2012), 9930–9942.

[113] D.E. Knuth, *The Art of Computer Programming*, Sorting and Searching, Reading, Mass.: Addison-Wesley, Vol. 3, 1973, 65–67, MR 0445948.

[114] J. Kondo, *Integral Equations*, Oxford Appl. Math. Comput. Sci. Ser. Oxford University Press, NewYork: The Clarendon Press, Kodansha, Ltd., Tokyo, 1991.

[115] J.C. Lagarias, Euler's constant: Euler's work and modern developments, *Bullet. (New Series) Am. Math. Soc.* 50(4) (2013), 527–628. Article electronically published on July 19, 2013.

[116] H. Lamb, On the diffraction of a solitary wave, *Proc. London Math. Soc.*, 8 (1910), 422.

[117] N. Laskin, Fractional Poisson process, *Commun. Nonlinear Sci. Numer. Simul.* 8 (2003), 201. N. Laskin, Some applications of the fractional Poisson probability distribution, *J. Math. Phys.* 50 (2009), 113513.

[118] H. Levine and J. Schwinger, On the theory of diffraction by an aperture in an infinite plane screen, I., *Phys. Rev.*, 74 (1948), 958–974.

[119] W.H. Louisell, *Quantum Statistical Properties of Radiation*, Canada: John Wiley & Sons Limited, 1973.

[120] J.L. López and P.J. Pagola, *Analytic formulas for the evaluation of the Pearcey integral*, arXiv:1601.03615 [mat.NA].

[121] E. Lucas, Theorie nouvelle des nombres de Bernoulli et d'Euler, *Comptes rendus del'Academie des Sciences (Paris)*, 83 (1876), 539–541; *Annali di Matematica pura ed applicata, Serie 2*, 8 (1877), 56–79.

[122] T. Mansour, Commutation Relations, Normal Ordering, and Stirling Numbers, Chapman and Hall/CRC, 2015.

[123] E. Majorana, *Nuovo Cimento*, 14 (1937), 171.

[124] M.G. Mittag-Leffler, "Une généralisation de l'intégrale de Laplace-Abel, *Comptes Rendus Hebdomadaires des Séances de l'Académie des Sciences*, 136 (1903), 537–539.

[125] P.A. Martin, *Harry Bateman: from Manchester to manuscript project*, www.ima.org.uk/.../mtapril10/_harry_bateman_from_manchester_to_manuscript_project.pdf.

[126] I. Mezo, Exponential generating function of hyper-harmonic numbers indexed by arithmetic progressions, *Cent. Eur. J. Math.*, 11(3) (2013), 931–939.

[127] V.H. Moll and C. H. Vignat, *On polynomials connected to powers of Bessel functions*, (2013), arXiv:1306.1224v1 [math-ph].

[128] R.M. Murray, Z. Li, and S.S. Sastry, *A Mathematical Introduction to Robotic Manipulation*, Boca Raton, FL: CRC Press, 1994.

[129] M. Naber, Time fractional Schrodinger equation, *J. Math. Phys.*, 45(8) (2004), 3339–3352.

[130] P.J. Nahin, *Oliver Heaviside: The Life, Work, and Times of an Electrical Genius of the Victorian Age*, Baltimore: Johns Hopkins University Press, 1988.

[131] N. Nielsen, Recherches sur les polynomes d'Hermite, Mathematisk-Fysiske Meddelelser, vol. 1, 1918, p. 79, Det. Kgl, Danske Videnskabernes Selskab.

[132] M.M. Nieto and D. Rodney Truax, Arbitrary-order Hermite generating functions for obtaining arbitrary-order coherent and squeezed states, *Phys. Lett. A* 208 (1995), 8–16.

[133] K.S. Nisar, S. R. Mondal, P. Agarwal, and M. Al-Dhaifallah, The umbral operator and the integration involving generalized Bessel-type function, *Open Math.*, 13 (2015), 426–435.

[134] K.B. Oldham and J. Spanier, The fractional calculus: Theory and applications of differentiation and integration to arbitrary order, *Math. Sci. Eng.*, 111 (1974).

[135] K.A. Olive *et al.*, Review of particle physics, (Particle Data Group), *Chin. Phys. C*, 38 (2014) and 2015 update, 090001.

[136] G. Pagnini and R.K. Saxena, *A note on the Voigt profile function*, arXiv:0805.2274 [math-ph].

[137] R.B. Paris and W.N.C. Sy, Influence of equilibrium shear flow along the magnetic field on the resistive tearing instability, *Phys. Fluids*, 26(10) (1983), 2966–2975.

[138] M.A. Pathan, The Multivariable Voigt functions and their representations, *Scientia, Series A: Math. Sci.*, 12 (2006), 9–15.

[139] M.A. Pathan, M. Kamarujjama, and M. Khursheed Alam, On multiindices and multivariables presentation of the Voigt functions, *J. Comput. Appl. Math.* 160(12) (2003), 251–257.

[140] G. Peacock, *A treatise on algebra*, 2nd ed., vol. II, Cambridge: J.J. Deighton, 1845.

[141] K.A. Penson, P. Blasiak, A. Horzela, A.I. Solomon, and G.H.E. Duchamp, Laguerre-type derivatives: Dobinski relations and combinatorial identities, *J. Math. Phys.*, 50(8) (2009), 083512.

[142] K. Penson and K. Gorska, Exact and explicit probability densities for one-sided Levy stable distributions, *Phys. Rev. Lett.*, (2010), 105-210604.

[143] H. Qin and R. Davidson, Courant-Snyder theory for coupled transverse dynamics of charged particles in electromagnetic focusing lattices, *Phys. Rev. Spec. Topics-Accelerators and Beams*, 12(6) (2009), doi:10.1103/PhysRevSTAB.12.064001.

[144] G. Raffelt and L. Stodolsky, Mixing of the photon with low-mass particles, *Phys. Rev. D* 37 (1988), 1237.

[145] P.E. Ricci, *Le funzioni Pseudo Iperboliche e Pseudo Trigonometriche*, Pubblicazioni dell'Istituto di Matematica Applicata, N 192, 1978.

[146] J. Riordan, *Introduction to Combinatorial Analysis*, Dover, 2002, pp. 85–86.

[147] S. Roman, The umbral calculus, in: *Pure and Applied Mathematics*, vol. 111, New York: Academic Press, Inc. [Harcourt Brace Jovanovich, Publishers], 1984.

[148] J.A. Rochowicz Jr, *Harmonic numbers: Insights, approximations and applications*, *Spreadsheets in Education (eJSiE)*, 8(2) (2015), Article 4.

[149] S. Roman, The theory of the umbral calculus, *J. Math. Anal. and Appl.*, 87 (1982), 58.

[150] S. Roman, *The Umbral Calculus*, New York: Dover Publications, 2005.

[151] S.M. Roman and G. Rota, The umbral calculus, *Adv. Math.*, 27(2) (1978), 95–188.

[152] D.C. Shaw, Perturbational results for diffraction of water-waves by nearly-vertical barriers, *IMA J. Appl. Math.*, 34 (1985), 99–117.

[153] I.M. Sheffer, Some properties of polynomial sets of type zero, *Duke Math. J.*, 5(3) (1939), 590–622.

[154] M.D. Schmidt, Zeta Series Generating Function Transformations Related to Polylogarithm Functions and the k-Order Harmonic Numbers, *Online J. Anal. Combin.*, (12), Article 2 (2017).

[155] I.N. Sneddon, *Fourier Transforms*, Courier Corporation, 1995.

[156] J. Sondow and E.W. Weisstein, *Harmonic number*, Math World A Wolfram Web Resource, http://mathworld.wolfram.com/Harmonic Number.html.

[157] H.M. Srivastava and L. Manocha, A treatise on generating functions, *Bull. Amer. Math. Soc. (N.S.)*, 19(1) (1988), 346–348.

[158] H.M. Srivastava and E.A. Miller, A unified presentation of the Voigt functions, *Astrophys. Space Sci.*, 135(1) (1987), 111–118, (ISSN 0004-640X).

[159] K. Steffen, *Fundamentals of Accelerator Optics*, ed. by S. Turner, CERN Accelerator School, Synchrotron Radiation and Free Electron Lasers, April 1989, (1990), pp. 90–93.

[160] J.F. Steffensen, *Interpolation*, Baltimore: The Williams & Wilkins Company, 1927.

[161] J.F. Steffensen, On the definition of the central factorial, *J. Inst. Actuaries*, 64 (1933), 165–168.

[162] J.F. Steffensen, The poweroid, an extension of the mathematical notion of power, *Acta Mathematica*, 73 (1941), 333–366.

[163] V. Strehl, Lacunary Laguerre series from a combinatorial perspective, *Séminaire Lotharingien de Combinatoire*, 76 (2017), Article B76c.

[164] J. Sun, J. Wang, and C. Wang, Orthonormalized eigenstates of cubic and higher powers of the annihilation operator, *Phys. Rev.* 44A (1991), 3369.

[165] A. Thorne, U. Litzén, and S. Johansson, *Spectrophysics: Principles and Applications*, Springer, 1999, ISBN 13: 9783540651178.

[166] F.G. Tricomi, *Funzioni Speciali*, Gheroni, 1959, pp. 408.

[167] V.V. Uchaikin and V.M. Zolotarevm, *Chance and Stability: Stable Distributions and their Applications*, Walter de Gruyter, 01 gen 1999.

[168] S. Vajda, *Fibonacci and Lucas Numbers and the Golden Section: Theory and Applications*, New York: Halsted, Press, 1989, 9–10, 18–20.

[169] J.V. Wehausen and E.V. Laitone, Surface Waves, *Handbuch der Physik*, 3(9) (1960), 446–778.

[170] H. Weyl, *The Theory of Groups and Quantum Mechanics*, Dover Pub. Inc, 1950.

[171] E.W. Weisstein, *Hermite number*, from MathWorld–A Wolfram Web Resource, http://mathworld.wolfram.com/HermiteNumber.html.

[172] E.W. Weisstein, *Legendre polynomial*. From MathWorld–A Wolfram Web Resource, http://mathworld.wolfram.com/about/author.html.

[173] E.W. Weisstein, *Mittag-Leffler Function*. From MathWorld-A Wolfram Web Resource, http://mathworld.wolfram.com/Mittag-Leffler Function.html.

[174] E.W. Weisstein, *Motzkin number*. From MathWorld-A Wolfram Web Resource, http://mathworld.wolfram.com/MotzkinNumber.html.

[175] E.W. Weisstein, *Padovan sequence*. From Wolfram MathWorld-A Wolfram Web Resource, mathworld.wolfram.com/PadovanSequence.html.

[176] E.W. Weisstein, *Parabolic Cylinder Function*. From MathWorld–A Wolfram Web Resource http://mathworld.wolfram.com/Parabolic CylinderFunction.html.

[177] E.W. Weisstein, *Perrin sequence*. From Wolfram MathWorld-A Wolfram Web Resource, mathworld.wolfram.com/ PerrinSequence .html.

[178] D.V. Widder, *The Heat Equation*, Academic Press, 1976.

[179] E.M. Wright, The Asymptotic Expansion of the Generalized Bessel Functions, *Proc. London Math. Soc.*, 2(38) (1935), 257–270.

[180] K. Zhukovsky and G. Dattoli, *Umbral methods, Combinatorial identities and harmonic numbers*, *Appl. Math.*, 1 (2011), 46.

Index